Units of the Royal Australian Air Force

This book is one of a series of publications produced for the 75th Anniversary of the Royal Australian Air Force. Principal sponsors of Anniversary activities are Ansett Australia, British Aerospace Australia Limited, McDonnell Douglas and Lockheed Martin Aeronautical Systems.

ANSETT AUSTRALIA
One of the world's great airlines.

BRITISH AEROSPACE
AUSTRALIA

LOCKHEED MARTIN
Aeronautical Systems

MCDONNELL
DOUGLAS

UNITS OF THE ROYAL AUSTRALIAN AIR FORCE
A CONCISE HISTORY

VOLUME 7
MAINTENANCE UNITS

Compiled by the RAAF Historical Section

An AGPS Press publication
Australian Government Publishing Service
Canberra

National Library of Australia Cataloguing-in-Publication data:

Units of the Royal Australian Air Force: a concise history.
Volume 7, Maintenance units.

ISBN 0 644 42798 1.
ISBN 0 644 42793 0 (set).

1. Australia. Royal Australian Air Force - History.
I. Australia. Royal Australian Air Force. Historical Section.

358.400994

Cover photograph: Installation and testing of Rolls Royce engines at 1 Aircraft Depot, 1945

Produced by the Australian Government Publishing Service

CONTENTS

ILLUSTRATIONS

All photographs appearing in this volume are RAAF official photographs unless otherwise identified.

INTRODUCTION

In November 1992 it was proposed that, as part of the RAAF 75th Anniversary celebrations to be held in 1996, a complete history of each unit of the RAAF should be prepared. The concept was to produce a concise history of each unit which would also serve as a basic reference for further research, and the realisation of that concept has resulted in a unique work. Other air forces have produced a history of all operational units, but no other air force has attempted to give a concise history of every unit which was an integral part of its organisation. An early decision was to produce 10 'generic' volumes, each pertinent to a given function of the Royal Australian Air Force. Every effort has been made to remain true to this concept, although it is conceded that some arbitrary decisions have been made as to which volume a unit would best fit into.

After much discussion, a standard format was established for each volume and requests circulated for volunteers to research and write the histories. Each history was to be based on the relevant Unit History Record and every care taken to ensure factual accuracy within a strict word limit. Only information up to November 1994 was to be included.

Over 110 volunteers, drawn from within the Service, ex-service organisations and interested civilians, forwarded pieces to the RAAF Historical Section, which undertook the collation of the manuscripts, final checking and liaison with the publishers. Except for editing for consistency of style the histories remain largely as they were written.

During World War II many crashed aircraft were salvaged, repaired and returned to service by maintenance crews working long hours, often under fire and with next to nothing in the way of technical facilities. Volume 7, Maintenance Units, tells the story of those technicians whose work began with the wood and fabric aircraft of the 1920s and has progressed to the high technology carbon fibres and complex electronics of a modern air force. It is these technicians who have upheld the RAAF's 75-year history of technical excellence.

ACKNOWLEDGMENTS

It would not have been possible to complete a project of this size in the time allowed without the dedication of many people. In particular the RAAF Historical Section wishes to acknowledge the contribution of the volunteers who prepared the individual histories.

Allen, J.
Angel, M.
Antonakus, A.
Ashworth, N.
Bain, R.
Barden, A.
Barnes, N.
Bartram, G.
Beck, J.
Bennett, J.
Betts, T.
Boydell, R.
Brassil, J.
Brownlee, M.
Buchanan, M.
Burton, J.
Cathcart, E.
Charles, A.
Christianson, R.
Clark, J.
Connell, P.
Culbert, J.
Davis, P.
Diamond, M.
Durant, M.
Dyne, G.
Eldridge, M.
Faigniez, M.
Ferguson, G.
Firkins, P.
Fisher, K.
Fitzgerald, T.
Flynn, K.
Forno, I.
Freeman, B.
Fulton, D.
Funch, C.
Givern, S.
Gordon, I.
Green, I.
Grey, A.
Grey, G.
Griggs, P.
Hall, R.
Harding, J.
Harvey, M.
Haynes, A.
Haywood, P.
Hind, C.
Hocking, G.
Holden, R.
Holder, D.
Jorgenson, M.
Kenyon, C.
Kerr, M.
Knight, M.
Lamble, B.
Lawson, S.
Lax, M.
Layton, A.
Maconachie, B.
Magraith, J.
Marshall, G.
Maslin, D.
McConnell, M.
McCormack, N.
McLean, D.
Meertens, H.
Menso, K.
Mitchell, A.
Moore, M.
Moore, W.
Morrisey, M.
Muir, P.
Naughton, D.
Newton, D.
Nicholson, J.
Nicholson, K.
O'Donnell, M.
Owen, S.
Parker, J.
Paterson, R.
Phillips, N.
Phillips, P.
Pinder, C.
Pinder, G.
Piper, R.
Robinson, G.
Roddy, R.
Rolland, D.
Rose, I.
Schofield, J.
Scott, D.
Sexton, P.
Sherman, E.
Simmonds, E.
Siviour, J.
Smith, F.
Smith, P.
Stevens, G.
Sullivan, L.
Telfer, J.
Thiel, O.
Tilley, G.
Toohey, J.
Walton, G.
Waud, C.
Weller, E.
Williams, N.
Wills, R.
Wilson, D.

The RAAF Historical Section also wishes to thank the staff of RAAF Museum, Point Cook, the Central Photographic Establishment, Laverton, and AGPS Press.

1 AIRCRAFT DEPOT

Providemus (We Foresee)

CHRONOLOGY

21 July 1921	—formed at Point Cook, Vic.
1 March 1926	—re-located to Laverton, Vic.
2 December 1994	—Depot disbanded

NARRATIVE

Although 1 Aircraft Depot had its official beginning on 21 July 1921, a post detachment of the Central Flying School, Point Cook, Victoria, performed the duties of a depot from May 1920. The detachment handled large quantities of aircraft, motor transport and associated equipment presented to Australia by the British Government after World War I. The detachment, plus the Motor Transport Repair Section at North Fitzroy, Victoria, was formally established as 1 Aircraft Depot on 21 July 1921, with the role of assembling, maintaining and overhauling equipment for the newly formed Royal Australian Air Force. As increased numbers of aircraft and equipment arrived, better facilities and centralisation of support were required. Hence the Depot was re-located to Laverton, Victoria, on 1 March 1926 under the name of 1 Aircraft Depot, RAAF Station with a strength of eight officers and 77 personnel.

Support for government and other research activities also commenced. A Northern Survey Flight was formed in August 1926 with one DH-50 and two DH-9 aircraft. The expedition of six personnel, led by Group Captain R. Williams, inspected air route facilities and potential airfield sites around Australia. On 2 August 1927 a Papuan Survey Flight was formed, consisting of two Seagull amphibian aircraft plus six personnel led by Flight Lieutenant E.C. Wackett. This expedition performed reconnaissance of the coastal districts of Papua and mandated territories as far north as Aitape, and returned on 19 January 1928. Also during those years the Depot continued to grow as workshops, offices, stores, medical/dental facilities, accommodation and messes were erected.

During the 1920s and 1930s 1 Aircraft Depot functioned as a receiving unit for the RAAF re-equipment programs, which included assembling and flight testing new aircraft prior to disbursement to flying units. These aircraft included Westland Wapiti, Bristol Bulldog, Hawker Demon, de Havilland Rapide, Avro Cadet Trainer, Avro Anson and NA-16s. During this period the strength of the Depot increased to 350 personnel.

During September 1934 a Depot detachment provided assistance at Charleville, Queensland, for the Melbourne Centenary Air Race and on the 18 September 1935 a northern Australia geographical and geological survey, led by Sir Herbert Gepp, departed from Laverton in a de Havilland Rapide aircraft piloted by Squadron Leader G. Jones.

In September 1939 1 Aircraft Depot mobilised following the outbreak of World War II. All personnel lived in barracks as weekends and overtime were worked. Increased aircraft deliveries for the war effort included Wirraways, Tiger Moth trainers, Hudson, Hurricane and Kittyhawk machines. After assembly and test, ferry flights as far afield as Townsville, Queensland, and Port Moresby were carried out. During the War years the Depot also serviced operational aircraft and equipment. This included aircraft instrument repair, armaments installation and overhaul of several engine types including Cheetah, Wasp, Merlin and Kestrel.

During 1942 the Depot also assisted with servicing and repair of United States Army Air Corps B-17 Flying Fortresses which had sustained battle damage against Japanese forces. Considerable scientific research and aircraft development work was also provided during the War. Modification trials were carried out with Spitfire, Boomerang and Wackett Woomera aircraft, plus comparative trials between Zero, Spitfire, Boomerang, Kittyhawk and Buffalo aircraft.

1 Aircraft Depot reached peak strength of 2339 personnel in January 1945 and demobilised in late 1945. The first RAAF jet aircraft were introduced via 1 Aircraft Depot during 1946.

Aircraft maintenance continued after the War until 1961 when RAAF restructuring resulted in transfer of aircraft work to 2 and 3 Aircraft

Depots. During its aircraft maintenance history 1 Aircraft Depot maintained a total of 42 types of aircraft. Aircraft engine overhauls continued and new work began in installation and maintenance of telecommunications equipment.

In 1968 aircraft engine work was also discontinued and the Depot was re-organised to take on a new role. Work programs were then centred on ground-based equipment. During the 1970s contributions to development, manufacture, installation, technician training and maintenance support to the RAAF grew rapidly in the fields of air traffic control, navigation, telecommunications, vehicles and specialist ground support equipment. This work introduced and maintained numerous new technologies for the RAAF, directly providing much equipment to every major base in Australia plus defence co-operation areas and other government agencies.

On 6 November 1981 1 Aircraft Depot was presented a Queen's Colour. During the 1980s the Depot provided considerable prototype and modification production support including components for Iroquois, F-111, Macchi and F/A-18 Hornet aircraft. New technologies introduced to the RAAF via the Depot included electroplating, printed circuit board production and optical fibre technology. By September 1982 Depot strength had reduced to 235.

Defence policies resulting in the technical trades restructure brought about some reduction of the Depot's trade capability and increasing reliance on civilian contractor support from 1986 onwards. Further re-organisation of the RAAF from 1990 onwards meant that 1 Aircraft Depot functions were devolved to user units or transferred totally to civilian support. Hence, on 2 December 1994 1 Aircraft Depot was disbanded. At the date of disbandment 1 Aircraft Depot was the longest continuous serving unit of the Royal Australian Air Force, and the longest continuous serving maintenance depot of any air force in the world.

Commanding Officers

Detachment of Central Flying School

May 1920 —Flight Lieutenant H.W. Wrigley

1 Aircraft Depot

July 1921 —Squadron Leader W.H. Anderson

February 1922 —Squadron Leader A. Murray-Jones

July 1922	—Flight Lieutenant W.A. Coates
February 1923	—Squadron Leader T.R. Marsden
May 1924	—Flight Lieutenant E.J. Howells
December 1925	—Flight Lieutenant R. Christie
March 1926	—Wing Commander P.A. McBain
January 1929	—Squadron Leader J.R. Bell
April 1929	—Wing Commander W.H. Anderson
August 1929	—Wing Commander A.T. Cole
December 1932	—Squadron Leader R. Christie
January 1933	—Squadron Leader A.W. Murphy
February 1933	—Wing Commander F.H. McNamara
April 1936	—Wing Commander F.W.F. Lukis
January 1938	—Wing Commander A.W. Murphy
May 1940	—Wing Commander G.E. Douglas
July 1942	—Wing Commander J.C. Stevenson
December 1942	—Wing Commander J.M. Lerew
November 1943	—Wing Commander S.DeB. Griffiths
October 1945	—Group Captain C.McK. Henry
November 1949	—Wing Commander K.P. Connolly
July 1951	—Wing Commander E.F. Easterbrook
May 1953	—Group Captain E.V. Millett
October 1954	—Wing Commander A.T. Fay
June 1955	—Group Captain J.W.C. Black
March 1956	—Wing Commander R.W.T. Ayre
March 1957	—Group Captain J.P. Godsell
January 1959	—Wing Commander R.W.T. Ayre
June 1959	—Group Captain A.G. Pither
September 1959	—Wing Commander R.W.T. Ayre
January 1960	—Wing Commander A.G. Pither
June 1961	—Wing Commander F.M. Timms
September 1962	—Wing Commander J.R. Lavers
February 1966	—Wing Commander D.M. Furler
January 1968	—Wing Commander R.K. Starkie
June 1968	—Wing Commander H.J. Gent
September 1968	—Squadron Leader R.R. Prowse
May 1969	—Wing Commander D.C. Mazlin
November 1971	—Wing Commander J.G. Swales
January 1974	—Wing Commander G. Grantham
June 1974	—Wing Commander J.D.F. Philip

January 1977	—Wing Commander C.R.A. Ely
February 1979	—Squadron Leader P.J. Winyard
March 1979	—Wing Commander D.K. Leslie
January 1982	—Wing Commander L.J. Crowley
January 1984	—Wing Commander T.J. Kelly
June 1985	—Wing Commander R.I. Gretton
December 1987	—Wing Commander P.J. McDonald
January 1990	—Wing Commander T.G. Smith
January 1993	—Wing Commander M.W. Shaw

2 AIRCRAFT DEPOT

Dependable

Chronology

1 January 1936	—formed at Richmond, NSW
1 July 1992	—Depot disbanded

Narrative

2 Aircraft Depot was formed on 1 January 1936 at RAAF Station Richmond, New South Wales. Personnel of 3 Squadron assisted with erection of bin storage and temporary office areas. Wing Commander R. Christie was appointed temporary Commanding Officer.

During April 1936 workshop equipment from 3 Squadron was taken over together with reconditioning work on War Memorial aircraft. The first aircraft to be completely repaired was a Demon, completed in October 1936. The following months Demons were received from overseas for assembly. During December 1936 Seagull aircraft were road transported ex-ship from Sydney for assembly. On 14 December 1936 recruit training commenced, with 18 airmen posted into Technical Squadron.

In August 1937 Anson aircraft started to arrive from overseas and the working area resources of the unit were overtaxed. This was slightly relieved when Depot headquarters and stores office personnel shifted into their new headquarters building on 17 February 1938. A new Engine Repair Sector building was started in June 1938.

The advent of the Tiger Moth (first acceptance flight at Mascot, 16 May 1939) and the commencement of construction of the Wirraway at Commonwealth Aircraft Corporation, caused an increase in workloads at the Depot.

With the outbreak of World War II, 2 Aircraft Depot had grown from a nucleus of one officer and 14 airmen to a strength of 11 officers and 297 airmen. During the early period of the War, assembly of Demons, Ansons and Seagulls continued. Tiger Moths received from de Havillands were test flown before being issued to training establishments.

World War II introduced a continuous workload for all sections—armament, erection, radar, aircraft repair, motor transport repair, machine shop, metal work, engine repair, W/T repair, instrument, parachute and electrical repair. Difficult but valuable work was carried out by airmen in salvaging crashed aircraft, some of which were to fly again, while others were stripped for useful components.

During April 1944 the unit erected, modified, operationally fitted or repaired seven Ansons, 16 Austers, two Beaufighters, six Beauforts, three Kittyhawks, 10 Spitfires, four Vengeance, four Venturas, one Mosquito, two Hudsons and a Douglas aircraft. Engines despatched after overhaul or repair included 30 Gipsy Major, two Single Row Cyclone and one Double Row Wasp.

For a period after the cessation of hostilities the flow of work was disrupted by the loss of technically trained men on demobilisation. Work entailed storage and inhibiting of equipment, ferrying of aircraft for storage and the shipment of Mustangs to Japan for use by the occupation force. In March 1947 unit strength was down to 227 against an establishment of 645. In September 1947 Engine Repair Squadron completed 14 engines for the Antarctic Expedition.

This period saw a number of firsts for the Depot. The first jet aircraft received by 2 Aircraft Depot was Vampire A78-1 received from de Havillands on 9 June 1948. In July 1948 15 Spitfires and four Vultee Vengeance aircraft were made ready for handover to the Royal Australian Navy. Assembly of the first helicopters in New South Wales was completed by the unit in June 1951. The first Canberra bomber arrived at Richmond, New South Wales, in August 1951 flown by Wing Commander Cuming. By 1953 strength was up to 617 against an establishment of 663.

Aircraft issued from work during 1955 comprised Vampires and Neptunes, while Engine Squadron continued to overhaul aircraft engines. Army Inspection Directorate conducted Army parachute dropping tests with 2 Aircraft Depot in November 1956.

In January 1959 Wing Commander K.P. Connolly (Commander of 478 Squadron) visited the unit regarding technical assistance for movement of Sabre aircraft to Malaysia. Trial loading of Sabre aircraft into Hercules aircraft was successful and the aircraft were flown to Malaysia. Later in the year a work party went to Rathmines, New South Wales, where they dismantled Catalina 'Frigate Bird', made famous by Captain P.G. Taylor on his trans-Pacific flight.

In the late 1950s 2 Aircraft Depot began preparations for the arrival of the C-130A Hercules aircraft for major maintenance, and by the early 1960s a regular program of servicing and modifications had developed. A government decision to involve civil industry in Depot level maintenance saw a winding back of 2 Aircraft Depot's involvement in aircraft engine overhauls. Later, when more Hercules aircraft came on RAAF strength, some of the Hercules airframe workload was taken over by Qantas. Nevertheless the Depot established a reputation for excellence during the Hercules wing replank program including a commendation from the Lockheed Aircraft Corporation in the United States of America.

Maintenance programs for the Caribou, Orion and VIP aircraft were also developed by 2 Aircraft Depot during the 1970s and 1980s. In mid-1980 the Depot acquired the capability to overhaul Mirage aircraft and performed significant structural upgrades right through until the Mirage was withdrawn from RAAF service.

On 24 March 1982 2 Aircraft Depot, in recognition of giving technical support to the RAAF for 46 years, was presented with a Queen's Colour by the Governor-General, Sir Zelman Cowen.

Over the years 2 Aircraft Depot has been responsible for conducting courses such as: parachute, fork lift operators, equipment assistants, supervision and management, precision measuring equipment laboratory and base calibration centre.

On 1 July 1992, 503 Wing assumed the functions of 2 Aircraft Depot (and other units) and is still active.

Commanding Officers

24 August 1936	—Wing Commander R.J. Brownell
2 September 1939	—Wing Commander R. Christie
20 November 1939	—Wing Commander H.B. Seekamp
6 April 1942	—Wing Commander F.I. Stevens
1 September 1942	—Wing Commander H.S. Owen
30 May 1943	—Group Captain S.F. Kearney
7 January 1946	—Group Captain E. Hey

19 September 1947	—Wing Commander C.R. Hackfath
7 June 1950	—Wing Commander A.T. Fay
14 November 1951	—Wing Commander H.H. Smith
7 April 1952	—Wing Commander P.S. Norris
6 August 1952	—Wing Commander J.A. Campbell
2 July 1956	—Wing Commander J.A. Weddell
13 June 1959	—Group Captain J.P. Godsell
20 July 1959	—Wing Commander J.E. Haigh
9 May 1960	—Group Captain H.H. Smith
29 January 1963	—Group Captain C.R. Taylor
13 December 1966	—Group Captain G.F. Ell
3 September 1970	—Group Captain J.F. Flannery
11 January 1971	—Group Captain A.W. Badman
23 July 1973	—Wing Commander R.A. Redfern
14 August 1973	—Group Captain L.P. Bek
24 January 1975	—Group Captain J.A. Turton
2 January 1976	—Group Captain A.R. Rundle
15 January 1979	—Group Captain R.A. Redfern
2 July 1982	—Group Captain P.J. Rushbridge
17 July 1984	—Group Captain G.E. Lynch
19 June 1985	—Group Captain J. McCombe
4 January 1988	—Group Captain R.C. Bennett
15 December 1989	—Group Captain G.E. Smith

3 AIRCRAFT DEPOT

Excel

CHRONOLOGY

16 March 1942	—formed at Amberley, Qld
1 August 1942	—Headquarters RAAF Station Amberley disbanded, 3 Aircraft Depot took control of station
30 June 1992	—Depot disbanded

NARRATIVE

3 Aircraft Depot was formed at RAAF Station Amberley, Queensland, on 16 March 1942, under the temporary command of Squadron Leader W.H. Nicholson.

The Depot's function was to erect and despatch aircraft. By the end of April 1942, 3 Aircraft Depot had erected 123 P-39 Airacobra fighter aircraft and 12 B-26 Glen Martin Marauder medium bomber aircraft were erected by 1 May 1942. In addition, a salvage section was formed to salvage aircraft from surrounding squadrons and aerodromes. The salvaged repairable items were then sent to 5 Aircraft Depot at RAAF Base Wagga Wagga. The Depot also carried out major inspections of Wirraway and Hudson aircraft for operational squadrons and the manufacture of mountings and fittings for forward guns for Wirraways.

On 1 August 1942, Headquarters RAAF Station Amberley was disbanded and 3 Aircraft Depot assumed responsibility for all aerodrome maintenance personnel and control of the whole station.

On 31 July 1945, Mosquito A52-502 swerved on take-off and struck a civil truck standing about 40 feet from the edge of the runway. Two civilians were injured—Mr J.M. Walsh died of shock and multiple injuries and Mr E. Jensen was admitted to Ipswich District Hospital with minor injuries.

On 13 June 1950, Field Marshall Sir William Slim, Chief of the Imperial General Staff, arrived at RAAF Station Amberley and inspected the Army guard of honour. He was met on arrival by the General Officer Commanding Northern Command, the Commanding Officer of RAAF Station Amberley and senior Army and RAAF officers.

On 10 September 1950, the station was opened for the day and 25 000 people visited the station and witnessed displays by Lincoln, Mosquito and Mustang aircraft. A display of technical and safety equipment was prepared in Hangar 76, and approximately 15 000 people viewed the display.

On 14 August 1951, the Commanding Officer, Wing Commander J.A. Campbell, attended the opening of the Queensland State Parliament. A guard of 100 personnel was inspected by the State Governor, His Excellency, Lieutenant General the Right Honourable Sir John D. Lavarack. This was the first occasion on which a RAAF guard of honour had been invited to participate at this function.

On 22 September 1951, approximately 15 000 people visited the station during an open day. Displays by Lincoln, Mustang, Wirraway, Vampire and Canberra aircraft were flown. The Canberra had flown from Melbourne to Brisbane in a record-breaking 90-minute flight.

The 1970s F-111Cs were maintained at 3 Aircraft Depot as well as Iroquois and Chinook helicopters and Canberra bombers. By the 1980s the Depot was maintaining F-111C and Iroquois aircraft. At this time the Depot comprised a maintenance management squadron, consisting of an aircraft equipment maintenance flight (aircraft accessories, electrical and instrument sections), aircraft maintenance flight (armament section and deseal/reseal section) and engine maintenance flight.

3 Aircraft Depot was disbanded on 30 June 1992. 501 Wing took over its functions, and is still active.

Commanding Officers

18 June 1942	—Wing Commander G.E. Douglas
November 1946	—Wing Commander W.I. Matson
14 March 1949	—Wing Commander J.A. Campbell
23 July 1952	—Wing Commander J. Hearnden

28 May 1957	—Wing Commander F.M. Timms
31 October 1960	—Group Captain E.L. Chapman
14 December 1964	—Group Captain G.F. Ell
9 December 1966	—Group Captain J.A. Rowland
28 January 1969	—Group Captain L.S. Compton
July 1971	—Group Captain K.W. Fraser
4 July 1973	—Group Captain G.C. Monkley
6 August 1976	—Group Captain R.N. Wade
19 December 1977	—Group Captain D.C. Mazlin
15 January 1981	—Group Captain D.A. Tidd
11 January 1983	—Group Captain M.J. McDougal
16 August 1985	—Group Captain M.J. Brennan
March 1988	—Group Captain N.P. Middleton
23 January 1990	—Group Captain C.A. Tyler

4 AIRCRAFT DEPOT

CHRONOLOGY

15 May 1942	—formed at Pearce, WA
30 May 1942	—moved to Boulder, WA
30 April 1946	—Depot closed

NARRATIVE

4 Aircraft Depot was formed at Pearce on 15 May 1942 with a cadre of one officer and five airmen. At the end of the month it moved to its 'permanent' home, the racecourse in Boulder, just south of Kalgoorlie. Over the following months hangars, barrack blocks and other facilities were built and the adjacent civil airfield was taken over. By the end of 1942 the Depot was ready to start its task of repairing and overhauling aircraft engines, airframes and ancillary aircraft equipment, mainly for RAAF units located in Western Australia. At that stage it had a strength of 17 officers and 206 airmen.

The first aircraft to be worked on were six Brewster Buffalos from 25 Squadron at Pearce which were in urgent need of work on their gun mounts. By the end of December 1942 the engine repair workshops had been completed and the first batch of Pratt and Whitney R1830 14-cylinder twin-row radial engines had arrived for overhaul. These engines, the overhaul of which was to be a major task for the Depot over the next three years, were used in United States Navy and Qantas Catalinas as well as RAAF Beaufort, Dakota and Boomerang aircraft. The Catalina engines were sent to Boulder by rail, while those from RAAF aircraft normally came with the aircraft.

By mid-1943 construction of the Depot was still only 75 per cent complete, but work on aircraft, engines and ancillary equipment was proceeding at a rapid rate. By then the Depot strength had grown to 26 officers, 664 airmen and 24 WAAAFs. The main aircraft being overhauled were Beauforts from 14 Squadron at Pearce.

The second half of 1943 saw the completion of the rest of the facilities and the build-up of the Depot strength to almost 1000. Work was well in hand overhauling Boomerangs from 85 Squadron at Guildford and Vultee Vengeances from 25 Squadron at Pearce, as well as 14 Squadron Beauforts. Other aircraft types that passed through the Depot during this period were Wirraways, Dakotas, Hudsons, Kittyhawks and a Lodestar. By January the following repair sections and shops were in full production: engine, propeller, instrument, bombsight, parachute, hydraulic, armament, radio, radar, sheet metal, machine shop, coppersmith, blacksmith, carpentry, fabric, welding, fitting and turning, paint shop, salt bath, electroplating and engine test bed. On the flying side there was a flight test section, air traffic control and tarmac servicing. This latter section was kept busy, not only with aircraft being tested and ferried to and from the Depot, but also aircraft transiting between Perth and eastern Australia.

In December 1943 the Depot received its first, and only, test and ferry pilot. Until that time this task had been carried out by squadron aircrew. The pilot's job was to test fly aircraft after repair or overhaul to ensure that it was fully serviceable for squadron service. Once three or so flights were completed, the aircraft was returned to its unit, either by the test pilot or squadron aircrew, and another aircraft was collected for ferry to the Depot.

Throughout 1944 work continued at a steady pace. The 'statistics' for 27 May 1944 reveal a typical workload: 18 aircraft undergoing major inspection, major modification or crash repair; 22 Pratt and Whitney R1830 engines being rebuilt; 28 magnetos, six propellers, 11 carburettors, 29 hydraulic pumps, 440 spark plugs and other equipment being rebuilt. Also in May, test flights were averaging 24 per month and ferry flights, by the Depot test pilot, six per month. The tarmac servicing team were handling an average of 80 transit aircraft per month.

Life in general for those posted to the Depot was good. The residents of the goldfields readily accepted the Depot and its personnel. This, together with the relaxed local leave rules, ease of travel into town, the abundance of recreational facilities and a plentiful beer supply (unlike in many other parts of Australia) resulted in very high morale and subsequent high level of performance and workmanship.

Towards the end of 1944, when 85 Squadron at Guildford was re-equipped with Spitfires, its Boomerangs were ferried to the Depot to be held in fully operational condition in the replenishment pool. This meant that they had to be flown once per week which, with some 20 aircraft, meant that the test pilot had to carry out up to five flights per day over the three months that the aircraft were held on strength. In March 1945 the majority were ferried out to 83 Squadron in Queensland.

Also at the end of 1944, when 25 Squadron at Pearce was re-equipped with Liberators, its Vultee Vengeances were ferried to the Depot for storage and subsequent disposal.

After mid-1945 the workload of the Depot started to fall off due to a general reduction of operational activity in Western Area and the departure of the United States Navy Catalinas from Perth. Then, after the Japanese surrender in mid-August, work was restricted to the repair and overhaul of currently held equipment. From then until April 1946, when the Depot closed and all facilities were handed over to a care and maintenance unit, there was a steady run down in personnel. In November 1945 the radio, radar, welding, blacksmith, coppersmith, fitting and turning, and electrical sections were closed, followed by the engine repair, airframe repair and general engineering sections in January 1946. The care and maintenance unit was closed in April 1947.

Commanding Officers

15 May 1942	—Flight Lieutenant E. Redman
8 June 1942	—Flying Officer G.H. Tournay
1 July 1942	—Squadron Leader L.J. Walton
15 January 1943	—Squadron Leader E.L. Chapman
7 February 1944	—Wing Commander W.H. Nicholson
25 October 1945	—Flight Lieutenant A.D. Saunders
20 February 1946	—Flight Lieutenant A.J. Wright

5 AIRCRAFT DEPOT

CHRONOLOGY

23 March 1942	—formed at Forest Hill, Wagga Wagga, NSW
22 February 1946	—Depot disbanded

NARRATIVE

5 Aircraft Depot was formed at Forest Hill, Wagga Wagga, New South Wales, on 23 March 1942, under the temporary command of Squadron Leader J.C. Stevenson. The unit shared the airfield with 2 Service Flying Training School and, later, 5 Operational Training Unit. The presence of the latter until November 1943, when it moved to Tocumwal, New South Wales, limited the full development of 5 Aircraft Depot. The operational training unit occupied two Bellman hangars and seven huts required for the expansion of the Depot, as well as placing pressure on messing and domestic facilities. At 31 August 1942, the strength of the unit stood at 42 officers and 1612 airmen. At the end of November 1942 the accommodation at the Depot was limited to 1785 personnel, and 320 were authorised to live in emergency quarters.

An advance party of nine officers and 192 airmen of the 4th Air Depot Group, United States Army Air Corps, under the command of Captain Milton B. McGuire, arrived at Wagga Wagga on 30 March 1942. The main body of 15 officers and 296 men arrived on 15 April. On 21 April Captain J.S. Shockley arrived with the 70th Bombardment Squadron, which was accommodated at the unit until 16 May 1942; the squadron returned for a month, commencing in August, to undertake practice on the Big Springs bombing range.

The Beaufighter-equipped 31 Squadron was formed at the base on 14 August 1942, and remained at Forest Hill until 3 November, when the rear party departed.

A fire in the electrical and instrument repair section on 27 October 1943 had severely curtailed activity to such a degree that repair work had to be diverted to Australian National Airlines, and a shortage of fitter IIAs and IIEs recorded during October/November also caused delay in Beaufort airframe and engine maintenance. Manpower problems were also reported in April 1944, because of the number of outward postings and the lack of time for skills to be passed to replacements, and the fact that the number of aircraft held on charge was not being compensated by an increase in technical staff. During May 1944 there were 200 inward/outward postings, and the unit held 281 aircraft on strength.

At 1 January 1944, the strength of the unit stood at 59 officers, 1687 airmen and 227 airwomen. An aircraft storage section was formed at Cootamundra, New South Wales, at the same time, initially with 10 Kittyhawks and three Beauforts. At the end of May 1944 there were 63 aircraft and one glider in storage at Cootamundra.

Despite the workload, there were occasions when personnel undertook extra duties. A party of four travelled to Ceduna, South Australia, to change the engines in three Vengeance aircraft during April 1944. On 16 July 1944, 14 airmen were despatched to Sydney for wharf labouring duties, under the direction of 1 Transport and Movement Office.

During its period of operation, 5 Aircraft Depot maintained and serviced Beaufort, Beaufighter, Kittyhawk, Hudson, Vengeance, Mitchell, Ventura, Anson, Wirraway and, once, a Catalina aircraft.

The visit of Wing Commander E.E. Easterbrock, the Commanding Officer of 2 Flying Boat Repair Depot, Rathmines, New South Wales, on 13 December 1945 was indicative of the future of the Depot. He inspected the workshop with the aim of having certain equipment allotted to his unit when 5 Aircraft Depot was disbanded.

Equipment from 1 Engineering School, destined for Ground Training School, arrived during 24 January 1946 and this marked the demise of 5 Aircraft Depot. On 22 February 1946, Care and Maintenance Unit, Wagga Wagga, under the command of Wing Commander F. Stiller, came into existence. On 18 March 1946 Squadron Leader W.J. Symons assumed command and the unit was disbanded in May.

On 1 December 1948, the unit was reactivated as Care and Maintenance Unit, Allonvale, and Flight Lieutenant H.C. Harman was appointed as the Commanding Officer. The main task of the unit was the segregation of equipment in Hangar 203, from where it was either allocated to other units or auctioned.

Disbandment of this care and maintenance unit was completed on 27 January 1950.

Commanding Officers

8 April 1942	—Wing Commander H.B. Seekamp
4 December 1943	—Wing Commander A.A. Saw
December 1944	—Group Captain H.G. Owen
October 1945	—Wing Commander C.R. Hackforth

6 AIRCRAFT DEPOT

CHRONOLOGY

14 October 1943	—formed at Oakey, Qld
1 February 1946	—Depot disbanded

NARRATIVE

6 Aircraft Depot was formed on 14 October 1943 at Oakey, Queensland, with the purpose of providing a maintenance facility which would ease the workload being undertaken by 3 Aircraft Depot at Amberley, Queensland.

The site, which was purchased from Mr F.H. Beutel and Mrs M.F. McPhee, Mr N. Smidt and Miss E.A. Walker on 13 November 1942, comprised 934 acres of land. The Civil Construction Corps had commenced work on facilities at the site on 15 March 1943, to enable 6 Aircraft Depot to commence operations on 23 December 1943.

The works facilities were housed in nine American hangars, which had originally been consigned to Singapore. However, with the increase in the quantity and types of equipment being maintained, it was necessary to expand the Depot, and on 1 February 1945 an additional 57 acres were procured, along with the powerhouse, workshop camp and sewerage easement areas.

The aerodrome consisted of two hard-surfaced runways and a perimeter taxiway from the western side of the airfield to the workshops. Both were constructed of consolidated gravel primed with fluxed bitumen, but propeller blast loosened the screening, damaging wooden propellers. On 15 December 1944, approval was given for the airstrip and taxiways to be sprayed with hot bitumen and sand.

The unit occupied its permanent camp site on 4 January 1944, and settled into its routine of erection, repair and inspection of aircraft and engines. The first aircraft inspected by the unit was Spitfire A58-27, which was received on 4 February 1944. Throughout its existence, the Depot erected and serviced Beaufort, Wirraway, Kittyhawk, Boomerang, Spitfire, Lancaster, Mustang and Norseman aircraft. In addition, the Depot erected Seafire aircraft for the Royal Navy and the staff of 6 Aircraft Depot overhauled Wasp, Merlin and Allison aircraft engines. Another very important aspect of the work undertaken by the Depot was the servicing of a large quantity of communications equipment, radio compass receivers, direction finding equipment and radio altimeters, some of which were from salvaged, or damaged, aircraft.

Personnel strength at 6 Aircraft Depot grew from 12 officers and airmen when the unit was raised, to a peak of 1339 on 1 October 1945.

At War's end, 6 Aircraft Depot was utilised to store surplus aircraft. At 10 December 1945, the aircraft held consisted of one Oxford, 40 Boomerangs, 50 Mustangs, 221 Kittyhawks and 232 Spitfires; 544 aircraft were waiting for disposal.

On 1 February 1946, 6 Aircraft Depot was disbanded and a care and maintenance unit (Care and Maintenance Unit, Oakey) was raised to maintain the aircraft and facilities and supervise the disposal of assets. Surplus equipment, such as machines and hydraulic presses used for repair work, were transferred to 3 Aircraft Depot at Amberley, and RAAF Station Oakey was categorised 'to be retained but not maintained'.

Commanding Officers

1 November 1943	—Wing Commander N.A. Adler
15 June 1945	—Wing Commander E.V. Millett
12 November 1945	—Squadron Leader W.W. Lalor

7 AIRCRAFT DEPOT

CHRONOLOGY

May 1942	—formed at Corowa, NSW
2 December 1942	—advance party to Tocumwal, NSW
14 February 1946	—facilities taken over by Care and Maintenance Unit, Tocumwal
June 1949	—last entry in Unit History Record

NARRATIVE

During May 1942, the Army impressed buildings at Albury, New South Wales, for the use of the RAAF at the local Corowa airfield. Plans were made for the expansion of the airfield and the dispersal of hangars and workshops. An advance party of two officers and 60 airmen, under the command of Flying Officer L.S. Gardiner, arrived at Corowa on 11 May 1942, and took up residence at the vacant Globe Hotel in Sangar Street. The Depot headquarters was established in the Literary Institute.

Flight Lieutenant F.D.A. Power was posted as administrative officer and temporary Commanding Officer on 27 May 1942. During June the overhaul of Twin Wasp engines commenced, and the non-commissioned officers moved from the Globe Hotel to quarters on 17 June.

Squadron Leader K.P. Connolly was appointed Commanding Officer of 7 Aircraft Depot on 17 July 1942, to face problems caused by the delay in the construction of the airfield and the slow arrival of maintenance equipment. The unit expanded slowly; on 14 September 1942, Flight Lieutenant A.H. Cottee arrived to establish the Aircraft Repair Squadron. By the end of October, the strength of the unit was 23 officers and 528 airmen.

Squadron Leader Connolly proceeded to Tocumwal on 1 December 1942 to check the route from Corowa to that site and the location of the unit there. The move was authorised on the following day, and an advance party, under the command of Flight Lieutenant Cottee, departed for Tocumwal that afternoon.

13 Aircraft Repair Depot was formed at Tocumwal on 5 December 1942, the nucleus of personnel being supplied by 7 Aircraft Depot.

The first large aircraft to be serviced at 7 Aircraft Depot was a Hudson, which arrived on 14 January 1943 for an engine change and inspection. Five days later Spitfires arrived for storage, and the unit strength was 1023 at the end of the month. During February, the propeller section and wireless maintenance shop were established, and the maintenance effort increased during March with the arrival of 20 Wackett Trainers for overhaul in addition to three Beauforts. Engines such as the Twin Row Wasp, Warner Scarab and Cheetah IX and X were overhauled.

Wasp maintenance was delayed when number 1 control room at the test stand in the engine repair squadron was totally destroyed by fire. This put number 1 and number 2 test cells out of action, and seriously affected the output of engines. Number 4 cell, which was being equipped to test R2800 engines, was fitted out to test the Wasp until the damage had been repaired.

From 13 June 1944, small groups of United States Navy personnel were attached to the unit for three-week periods for instruction in engine repair and maintenance.

Next month the first Liberator aircraft was serviced by the Depot.

On 19 August 1944, portions of the transport, stores and messing sections were transferred to station headquarters, Tocumwal. Even so, the unit strength in November 1945 stood at 1329.

On 14 February 1946, the facilities at the Depot were taken over by Care and Maintenance Unit, Tocumwal. The latter unit was to undertake work still in progress—the repair of two RAAF Liberators, the repair of the front fuselage of RAF Liberator EW634, maintenance of Netherlands East Indies aircraft, the completion of the engine repair program and the storage of aircraft. 398 aircraft, including Tiger Moth, Oxford, Wirraway and Liberator, were held at the end of February 1946.

Care and Maintenance Unit, Tocumwal, was commanded in turn by Flight Lieutenant S.T. Carter, Squadron Leaders J.A. O'Donnell, J.J. Mallon and A.H. Cottee. The final Commander was Flight Lieutenant T.W. Lynch, and the last entry in the Unit History Record was made in June 1949. At this stage equipment was being prepared for sale by the Commonwealth Disposal Commission.

Commanding Officers

17 July 1942	—Squadron Leader K.P. Connolly
August 1944	—Wing Commander F.W. Sexton
6 December 1945	—Wing Commander E.J. Brunckhurst

1 Aircraft Depot, Laverton, June 1926

1 Aircraft Depot headquarters, October 1972

1 Aircraft Depot Detachment 'B' disbandment parade at Tocumwal, New South Wales, 14 October 1960

Engine repairs at Richmond, circa 1930

Jet engine maintenance at 3 Aircraft Depot, early 1950s

3 Aircraft Depot hangar, August 1965

F-111C engine maintenance. L-R: Corporal Docksey, Sergeant Cassidy, Leading Aircraftman Hammond, Corporals Skinner, Connell, James and Marshall

14 Aircraft Repair Depot Airframe Section, November 1943

An aerial view of Flying Boat Maintenance Unit facilities at Bowen, Queensland

Salvage of aircraft entailed back-breaking work, as shown in this attempt to reclaim a Beaufort and...

...recover a burnt-out engine

Engine test cells at 5 Aircraft Depot, Wagga Wagga

5 Aircraft Depot engine workshop

Al fresco engine maintenance in Italy

Hangar built from local timber at Goodenough Island

1 AIRCRAFT PARK

CHRONOLOGY

29 April 1940	—formed within 1 Aircraft Depot, Laverton, Vic.
27 May 1940	—moved to Geelong, Vic.
30 January 1942	—re-located and placed under 1 Aircraft Depot control for handover to United States Army Air Corps
3 December 1942	—reformed at Geelong
19 October 1944	—unit disbanded

NARRATIVE

As the war effort grew, the Erection and Test Squadron of 1 Aircraft Depot, Laverton, Victoria, required expansion to cope with the ever increasing volume of aircraft arriving for assembly. Hence 1 Aircraft Park was formed as part of Erecting Section of 1 Aircraft Depot on 29 April 1940.

Flight Lieutenants H. Vatcher and H. Barnes, Royal Air Force, plus other RAF personnel reported to 1 Aircraft Depot and were subsequently posted to 1 Aircraft Park.

The unit moved to Geelong, Victoria, on 27 May 1940 and became a separate unit as of that date under the command of Flight Lieutenant H. Barnes, RAF. Four Fairy Battle aircraft had already arrived at Geelong awaiting assembly. The facilities consisted of an aircraft assembly hangar which was a converted large building commandeered from International Harvester, plus a cleared paddock used as a makeshift airfield.

During June 1940 13 Fairey Battle aircraft were delivered from 1 Aircraft Park to 1 Aircraft Depot for final touch-ups and ferry to user units. During the next 18 months the unit grew to a strength of approximately 130 personnel. A steady stream of aircraft was assembled; however corrosion was proving to be a problem particularly with the Battle aircraft which were being affected by the salt environment during sea freight from Britain.

At the end of 1941 a decision was made that the United States Army Air Corps would take over the unit. Hence 137 personnel, plus equipment and materials were transferred to 1 Aircraft Depot on

30 January 1942. During February 1942 Squadron Leader H.O.L. Digby resumed command of Aircraft Repair Squadron, 1 Aircraft Depot, after having handed over command of 1 Aircraft Park to the United States Army Air Corps. United States Army Air Corps aircraft of various types arrived and departed after completion of repairs and engine change, following damage sustained in action against Japanese forces operating in the near north.

On 3 December 1942 1 Aircraft Park was built up again at Geelong. By the end of that month 26 Airspeed Oxford aircraft were delivered to the unit from storage at Corpus Christie, Werribee, Victoria. Aircraftman Ronald Pike also arrived on duty in December 1942 as a freshly trained flight rigger. He recalls the airstrip as being undersized, very rough and the cause of regular aircraft mishaps.

Curtiss P-40 Kittyhawk aircraft were delivered and being assembled early in 1943. By early May 1943 initial shipments of six crated Spitfire aircraft were also being received for assembly. Work then progressed furiously with two shifts operating, placing severe strain on personnel. During June seven Spitfires were assembled and delivered to Production, Test and Ferry Flight of 1 Aircraft Depot.

1 Aircraft Park continued to provide aircraft assembly support to 1 Aircraft Depot until disbandment on 19 October 1944.

Commanding Officers

19 June 1940	—Flight Lieutenant J.M. Lerew
29 September 1940	—Squadron Leader H. Vatcher (RAF)
12 December 1941	—Squadron Leader H.O.L. Digby
3 December 1942	—Squadron Leader F.J.P. Woods
December 1943	—Squadron Leader W.W. Lalor

2 AIRCRAFT PARK

CHRONOLOGY

1 May 1940	—formed within 1 Aircraft Depot, Laverton, Vic.
19 December 1940	—re-located to Bankstown, NSW
28 March 1945	—unit disbanded

NARRATIVE

The nucleus of 2 Aircraft Park was formed within the Erecting Section, 1 Aircraft Depot, Laverton, Victoria, in May 1940 under the temporary command of Flying Officer Nicholson. The unit's role was to uncrate and assemble aircraft being received from overseas.

During May 1940 four Anson aircraft were assembled and delivered to the Erection and Test Squadron of 1 Aircraft Depot. Three more Ansons were assembled in June 1940.

On 19 December 1940 the unit was re-located to Bankstown, New South Wales, to provide aircraft assembly support to 2 Aircraft Depot.

On 30 September 1943 the unit received two horses for the purpose of patrolling the outer perimeter of the aerodrome. The patrolling was carried out by Mounted Service Police.

2 Aircraft Park was disbanded on 28 March 1945.

COMMANDING OFFICERS

23 June 1940	—Squadron Leader M.B. Allen
17 December 1941	—Squadron Leader N. Mulroney
10 March 1942	—Wing Commander A.D. Charlton
31 July 1942	—Squadron Leader R.S. Nickoll
12 December 1942	—Squadron Leader F.G. Huxley
3 May 1943	—Squadron Leader R.W.H. Harrison
4 May 1944	—Wing Commander F.I. Stevens

12 AIRCRAFT REPAIR DEPOT

CHRONOLOGY

1 May 1944	—formed at Werribee, Vic.
6 August 1944	—main body arrived at Darwin, NT
20 May 1946	—reduced to care and maintenance status
15 August 1946	—Depot disbanded

NARRATIVE

12 Aircraft Repair Depot was formed at Werribee, Victoria, on 1 May 1944, to repair and maintain B-24 Liberator aircraft in North-Western Area. Under instructions from the Commanding Officer, Wing Commander G.F. Ell, who was visiting North-Western Area Headquarters to arrange for the unit to transfer to the Northern Territory, Flying Officer A.L. Crear proceeded to Werribee to take charge of the nucleus of the unit.

The site at Werribee lacked messing and barrack facilities; meals and bedding had to be supplied by 1 Central Recovery Depot. On 7 May 1944 the orderly room, equipment and transport sections opened for business, with the first equipment for the Depot being delivered on 12 May.

On 24 May 1944, plans were made for the move of the unit to Darwin. The motor transport repair section was to commence its move on 7 July 1944, and the complete move was scheduled to be completed by 1 October 1944. Before the transfer, the strength of the former section had to be raised and, although the nucleus of the repair section had been posted in on 1 June, the actual departure had to be delayed as the officer commanding had not arrived. However, on 25 July, the motor transport repair section and 80 tons of stores departed for Darwin, where the unit was to take over a camp site vacated by 9 Works Maintenance Squadron.

Before the main body arrived in Darwin on 6 August 1944, the heavy equipment had to be prepared for the move to the tropics and packed. The men were ordered to undertake 'hardening' courses and 72 men commenced a course on 4 July 1944. The last 80 trained at Wonga Park for a week, commencing on 13 November 1944.

Unit members did participate in salvage work. On 17 August 1944, 10 personnel travelled to 1 Operational Training Unit to transport Beaufort A9-47 to 5 Aircraft Depot. On 29 August another party re-located A9-137 from the same site to 1 Central Recovery Depot and, on the 30th, a warrant officer and 18 servicemen travelled to Essendon to salvage a B-24 Liberator. This aircraft was converted to components and delivered to 7 Aircraft Depot, where a flight sergeant and three sergeants had been attached for a weeks training on B-24 aircraft during August.

As at 1 January 1945, the strength of 12 Aircraft Repair Depot was 930 officers and airmen, all working to establish the camp site and facilities near Darwin. On 17 January 1945, the *Mangola* arrived, and was unloaded by unit personnel. The activities of the Depot were not disturbed by the sighting of a Japanese reconnaissance aircraft flying over the town on 2 March 1945.

Arrangements had been made with the Commanding Officer of 49 American Air Depot for 12 Aircraft Repair Depot to have the use of six B-24 hulks, and another Liberator was salvaged from West Arm before the first Australian Liberators were issued to the Depot for maintenance on 17 March 1945. The first aircraft (A72-64) suffered from damage to the horizontal and vertical stabilisers; the second arrived during the next day for repair of bullet damage to the electrical wiring, instruments and structure. A72-64 was declared serviceable on 8 April 1945.

During June, 12 Aircraft Repair Depot accepted 36 aircraft for repair or maintenance, and 17 were delivered from its facilities. During the same month, discussions regarding the amalgamation of 12 and 14 Aircraft Repair Depots took place. The motor transport repair sections of both units were combined, but the end of the War curtailed the total merger.

By the end of August 1945, the policy was laid down that only top overhauls of engines would take place in the Depot, and that no more 600-hour inspections would be undertaken after the current servicing had been completed. The Depot would then undertake major maintenance for 82 and 85 Wings. However, at the end of November, RAAF Headquarters ordered that the unit complete the work in hand, and that only flying squadrons remain in the north.

The last aircraft serviced by 12 Aircraft Repair Depot (A72-76) departed for Tocumwal on 23 December 1945, and the Depot had reduced its strength to 234 officers and men. On 1 February 1946 orders were received to the effect that no more aircraft would be issued to the Depot for repair. At the end of February there were 130 men on strength; by the end of March 1946 no technical musterings remained.

On 20 May 1946 the unit was converted to a care and maintenance unit, and was disbanded on 15 August 1946.

COMMANDING OFFICERS

1 May 1944	—Wing Commander G.F. Ell
19 January 1946	—Squadron Leader T.H. Blake

13 AIRCRAFT REPAIR DEPOT

CHRONOLOGY

12 December 1942	—formed at Tocumwal, NSW
22 March 1943	—commenced move to Breddan, Qld
13 October 1947	—Depot disbanded

NARRATIVE

On 12 December 1942, 13 Aircraft Repair Depot was formed at Tocumwal, New South Wales, with Flight Lieutenant A.S. Johnson in temporary command. Squadron Leader R.R. Campbell was appointed as Commanding Officer on 5 January 1943, but was involved with the planning of the unit's move to its ultimate destination—Breddan, Queensland—leaving Flight Lieutenant A.G. Rethel in command pending Campbell's arrival at Tocumwal.

Flying Officer Oats led the advance party to Charters Towers, arriving on 21 February 1943. Permission was requested from Air Force Headquarters for the men to assist with the erection of hangars, buildings and facilities at Breddan. On 22 March, Flight Lieutenant Cottee led another 50 airmen to Breddan, where they joined the advance party undertaking the building task, to which was included the provision of electrical power. All these tasks were supervised by the Department of the Interior.

By the end of September 1943, the aircraft repair, general engineering, electrical repair, headquarters, motor transport repair, medical and dental, barracks, salvage and equipment sections had all moved to Breddan. The first record of aircraft being serviced was in October, when Kittyhawk, Tiger Moth, Beaufort and Spitfire aircraft were allotted to the Depot for repair. This development was not without incident: on 10 October a bushfire threatened the north-west of the Depot area, and tents were destroyed by fire on 14 October—the day after fire breaks had been burnt around important areas of the camp. Detachments were sent to Macrossan, where aircraft taken over at the 12 Repair and Salvage Unit site were serviced, and to Cooktown. Another detachment was established at Garbutt, Townsville, employed in the erection of Kittyhawk aircraft.

The body of men who remained at Tocumwal had not been idle. In addition to preparing equipment for shipment north, Flying Officer

W.R. Sleebs and party salvaged Beaufort A9-282 from 1 Operational Training Unit and delivered it to 7 Aircraft Depot on 1 June 1943.

By the end of October, the strength of the Depot was 991 officers and men. Two mobile salvage sections were formed, one marine and the other land, to operate in an area bound by Karumba, Cooktown, Rockhampton and Cloncurry.

These developments were marred by the destruction of the engine test stand on 31 December 1943, and the death of Leading Aircraftman W.G. Gadd, a passenger in Beaufort A9-452 which crash-landed five miles north of Fanning on 21 March 1944.

During April 1944, physical and recreation training was instituted, taking place for two hours each Saturday morning. Three football fields, six tennis courts and a rifle range were constructed for the use of members of the Depot, and sporting teams played against neighbouring units and civilian sides.

The unit reached its peak strength of 1020 on 30 September 1945. This figure included members of the WAAAF, the first four being posted to the Depot on 6 April 1945.

Leading Aircraftman B.D. Greene and Corporal J.W. Sutherland were killed in separate motor vehicle accidents at Charters Towers and Cloncurry on 26 March and 14 April 1945, respectively.

In July 1945 a Liberator arrived from Garbutt for storage, and the strength of the unit decreased markedly. When the unit was disbanded on 13 October 1947, its strength was 16 officers and men.

Commanding Officers

5 January 1943	—Squadron Leader R.R. Campbell
13 October 1943	—Wing Commander T.A. Cummings
1 February 1945	—Wing Commander W.B. Curtain
March 1945	—Squadron Leader J.W. Turton
July 1945	—Flying Officer H.T. Bell
9 April 1947	—Flight Lieutenant J.J. Rhyder

14 AIRCRAFT REPAIR DEPOT

CHRONOLOGY

25 September 1942	—raised as 7 Airframe Repair Depot at Ascot Vale, Vic.
30 October 1942	—re-named 14 Aircraft Repair Depot
8 February 1943	—commenced move to Gorrie, NT
20 October 1945	—Depot closed
11 March 1946	—recorded at Pearce, WA
31 July 1946	—Depot disbanded

NARRATIVE

Flight Lieutenant T.B. Philcox raised 7 Airframe Repair Depot at Ascot Vale, Victoria, on 25 September 1942. By 30 October, the unit had four officers and seven airmen on strength, and was re-named 14 Aircraft Repair Depot.

On 5 November 1942, Flying Officer W.W. Lalor was posted to the operational base at Birdum, Northern Territory, as the Commanding Officer of the Birdum detachment. On 8 January 1943, Squadron Leader F.W. Sexton travelled to Gorrie, Northern Territory, to inspect the proposed site for the unit, and returned to Melbourne on 18 January 1943.

At the end of January 14 Aircraft Repair Depot consisted of 13 officers and 175 men.

On 8 February 1943, Flight Lieutenant T.B. Philcox and an advance party of one other officer and 71 airmen departed for the Northern Territory, arriving at Gorrie on 18 February. This was the first of seven groups which undertook the eight-day trip: 224 officers and men departed on 8 March, 196 on 30 March, 304 on 11 April, 123 on 27 April and 45 on 14 May 1943. The final detachment departed from Ascot Vale on 22 September 1943. At the end of May, the strength of 14 Aircraft Repair Depot had grown to 959.

During June, the Depot was established and the first Spitfire (LZ848) arrived on 25 July 1943, starting a steady stream of the English fighters from 54 and 452 Squadrons. The Depot later undertook the maintenance of many aircraft types, including Mitchell, Beaufighter and Hudson. Salvage duties were also undertaken and during February 1944 the section at Darwin Civil Aerodrome became officially known as 14 Aircraft Repair Depot Forward, and aircraft were salvaged from as far afield as Tennant Creek.

The 'wet' season commenced in October, and 263 points of rain were recorded on the afternoon and evening of 17 December 1943. The 'wet' hampered work, but not the small arms training which commenced at Mataranka during September and the training of aerodrome defence instructors.

Service training was complemented by sport and recreational activities. The Airmen's Recreation Centre was opened on 27 November 1943, and this was the venue for a symphony concert party performance during February 1944. On 22 August 1943, a sports meeting was held at which all RAAF units south of Pine Creek participated to select a team to meet RAAF units north of that site. 14 Aircraft Repair Depot obtained 27 representatives in the squad which competed at Adelaide River on 4 September 1943; the result of the competition is not recorded. During September 1944, Sir Donald Cameron and a boxing and wrestling party gave competitive and exhibition matches at the unit.

Tragedy shook the Depot on 21 April 1944 when Flying Officer F.W. Miller, who had been detached to 4 Repair and Salvage Unit, died of injuries received when an oxygen unit exploded at that unit. Another unit member, Leading Aircraftman J.C. Hoffman, was killed instantly in the same incident.

During July 1945 it was decided that no further work would be accepted by the Depot, as it was planned to amalgamate it with 12 Aircraft Repair Depot at Darwin. During August, the motor transport repair and engine repair sections departed from Darwin. Arrangements for the amalgamation proceeded throughout August, broken by the celebrations to mark the Armistice on 15 August. The only misfortune was that suffered by the adjutant, Flying Officer J.G.M. McLauchlan, who broke his leg when he slipped over an empty beer bottle.

All the messes closed in September 1945. The aircraft held at the unit departed, and welfare, sporting and entertainment activity ceased. The Depot ceased operation on 22 September 1945, and was officially closed on 20 October 1945. In the period 1–19 October, all stores and equipment were transferred to 9 Stores Depot.

14 Aircraft Repair Depot still existed, and was active at Pearce, Western Australia, during March 1946.

On 18 July 1946, just before the unit disbanded, Corporals C.W. Chester and P.J. Wall and Leading Aircraftman R.A. Powell were killed in a vehicle accident on the Perth Road, three miles from Pearce.

During July, the unit occupied a vacant hangar and was disbanded on 31 July 1946. The Commanding Officer, Flight Lieutenant W. Morton, became the station engineering officer, and the nucleus of the unit formed the Pearce station workshop.

COMMANDING OFFICERS

5 January 1943	—Squadron Leader F.W. Sexton
3 June 1944	—Wing Commander C.R. Hackfath
1 August 1945	—Wing Commander J.S. Barling
11 March 1946	—Squadron Leader J.A. Campbell
22 March 1946	—Flight Lieutenant W. Morton
3 April 1946	—Squadron Leader P. Neale
24 June 1946	—Flight Lieutenant W. Morton

15 AIRCRAFT REPAIR DEPOT

CHRONOLOGY

7 March 1943	—officially formed at Port Moresby
February 1946	—reduced to care and maintenance basis
20 June 1946	—ceased to function
19 July 1946	—Depot disbanded

NARRATIVE

Prior to the official formation of 15 Aircraft Repair Depot at Port Moresby by 9 Operational Group on 7 March 1943, records are scarce. It is recorded that on 21 December 1942 Sergeant R.J. Anderson and party returned from Myola, where they had inspected Ford and Stinson aircraft. On the 27th, Sergeant R. Couper and another party arrived at Port Moresby after salvaging Stinson and Moth aircraft at Otamatu and Hula, as well as inspecting a Moth located at Cape Rodney.

As at 31 January 1943, the unit operated detachments at Gurney, Milne Bay (54 personnel), Buna (six personnel) and Wards Strip (86 personnel). Overall, the strength of the Depot was 426, with its headquarters located in the Port Moresby town area. However, on 28 February, the headquarters moved to Wards Strip, where 15 Aircraft Repair Depot facilities were developed.

The detachment at Milne Bay was closed with the advent of 10 Repair and Salvage Unit in the area. Sergeant Harrington and 32 men arrived back from Milne Bay on 21 March 1943.

April 1943 was not a month for airmen newly posted to the Depot. On the 12th, 25 airmen arrived on posting, and their arrival coincided with a raid of 43 enemy bombers and fighters. The dispersal bays and works area were bombed, without damage to the Depot facilities. Ten airmen, who had been posted to 15 Aircraft Repair Depot, were passengers on a flying boat which crashed at sea on 22 April. Three airmen were drowned and the seven survivors of the Depot contingent were rescued after being suspended from life jackets in the water for 18 hours.

During November 1943, an Aircraft Reserve Pool was formed at the Depot, to supply replacement aircraft to the operational squadrons flying over New Guinea. This remained in existence until January 1944, when, as a result of a re-organisation of the Depot, the pool was taken over by the maintenance section. Aircraft types which were maintained by the Depot included Beaufort, Kittyhawk, Vengeance, Wirraway, Boomerang, Boston, Beaufighter, Dragon and Spitfire aircraft. One of the more unusual events took place on 8 June 1944, when Flying Officer M.G. McSpedden ferried a Lightning, which had been taken over by the RAAF from the United States 5th Air Force, to Amberley.

On 24 June 1944, 15 Aircraft Repair Depot took over the responsibility of Wards Strip from 42 Operational Base Unit. In the same month, two WAAAF nurses arrived on temporary duty—the first females to serve with the unit. The Depot's assumption of responsibility for Wards Strip did not prevent aircraft maintenance tasks from being undertaken away from the Port Moresby area. On 8 July 1944, 12 airmen travelled to Cape Gloucester to complete repairs to aircraft following the move of 12 Repair and Salvage Unit from the area.

On 4 September 1944, a Wirraway and a Tiger Moth aircraft were issued to the unit for communications tasks. On 16 September, at the behest of the Army, Squadron Leader A.J.A. O'Donnell instituted a mail run to Yule Island, using the Tiger Moth.

During 21 June 1945, Squadron Leader O'Donnell and the unit doctor, Flight Lieutenant Roberts, responded to a request from the Australian New Guinea Administrative Unit (ANGAU) to evacuate a Captain Chester from Arouna Plantation. Roberts treated the patient at Arouna before he was flown to Wards Strip. On the following day, Chester's wife was flown out to be with her husband but the mercy dash was to no avail, as Chester died on 23 June.

The end of hostilities marked a rapid deterioration in unit strength. During February 1946, the unit was in the process of being reduced to care and maintenance basis. In April, the strength had fallen to 16, all ranks. During the month following the strength was down to four, and the unit ceased to function on 20 June 1946.

15 Aircraft Repair Depot was officially disbanded on 19 July 1946.

Commanding Officers

23 January 1943	—Squadron Leader W.R. Lowrie
17 February 1944	—Wing Commander E.V. Millett
19 November 1944	—Wing Commander P.S. Kennedy
2 March 1946	—Flying Officer C.R. Bagley

1 CENTRAL RECOVERY DEPOT

CHRONOLOGY

1 November 1943	—formed at Werribee, Vic.
21 February 1946	—last entry in Unit History Record

NARRATIVE

1 Central Recovery Depot was formed at Werribee, Victoria, on 1 November 1943, and Flight Lieutenant C.T.P. Jones was appointed as the Commanding Officer. Two days later hangars and workshops were taken over from the RAAF base at Laverton, as were the kitchens and huts at the 1 Service Flying Training School satellite airfield. On the same day, the Salvage Section from Point Cook moved to Werribee.

Work commenced on the sleeping accommodation, medical hut, recreation room and messing hut. Such was the shortage of supplies that these facilities were constructed from discarded aeroplane storage cases.

By February 1944, the unit had developed to the extent that it produced 1200 ball and roller bearings, stripped 34 batteries and reclaimed 20 more. The stripping of bearings and batteries and the reconditioning of the latter was the major reported activity of the unit. However, the unit was involved in the recovery of all types of useable metal and instruments from salvaged aircraft. For example, corrosion was removed from piston rings and these took their place, where applicable, as spare parts. In April 1944, there were 123 serviceable Fairey Battle aircraft held at the unit, awaiting a decision as to their ultimate fate.

The salvage parties ranged far and wide. Anson aircraft were recovered from Yanakie and Essendon; Vengeance aircraft from Mount Disappointment, Nhill and Laverton; Beaufort aircraft from Cressy, Bairnsdale and Mallacoota. In addition, the unit salvaged Spitfire, Boomerang, Ventura, Mitchell, Oxford, Liberator, Tiger Moth, Wackett Trainer, Hudson and Demon aircraft for conversion to spare parts and scrap.

A non-technical salvage depot was established at 11 Nott Street, Port Melbourne, which accepted salvaged stores from contractors. During the period February 1944–August 1945, recoveries to the total value of £257 341 ($514 682) were re-used by the Air Force.

In October 1945, there were 473 servicemen employed at 1 Central Recovery Depot. In February 1946, the date of the last entry in the Unit History Record, the strength had fallen to 205.

Commanding Officers

1 November 1943	—Flight Lieutenant C.T.P. Jones
3 January 1944	—Squadron Leader E.W.J. Eastgate
17 February 1944	—Flight Lieutenant C.T.P. Jones
11 January 1945	—Squadron Leader H.H. Powell

2 CENTRAL RECOVERY DEPOT

CHRONOLOGY

15 January 1944	—formed at Richmond, NSW
30 June 1946	—last entry in Unit History Record

NARRATIVE

When Flight Lieutenant W. Lewis established 2 Central Recovery Depot at Richmond on 15 January 1944, the strength of the unit was two officers and 80 other ranks. By the end of the month, the unit had taken over the staff of the salvage sections at 5 Aircraft Depot, Wagga Wagga, 2 Stores Depot and 2 Flying Boat Repair Depot at Rathmines. The former became 2 Central Recovery Depot Detachment 'B', under the command of Flight Lieutenant E.A. Bray, on 2 June 1944, and, at the end of January 1944, the strength of the unit was four officers and 128 other ranks.

Salvage crews travelled to Nowra, Bankstown, Newcastle, Benalla, Berrigan, Williamtown, Camden, Old Bar, Coffs Harbour, Teralga, Goulburn, Mount Druitt and Kempsey to recover Beaufort, Avro Cadet, Kittyhawk, Dakota, Tiger Moth, Boomerang, Ventura, Beaufighter, Seagull and Mosquito aircraft. Not all of these tasks were straightforward. On 2 April 1944, Warrant Officer Reddaway was slightly injured when a heavy tender overturned during the salvage of a 15 Squadron Beaufort. An identical aircraft created problems during a recovery from Coffs Harbour during December 1944. Warrant Officer Hanna and his crew were forced to construct a special tool to remove the firing pistols from the depth charges in the aircraft before unloading the potentially deadly cargo.

A non-technical recovery section, Detachment 'A', was established at Mascot, under the command of Flying Officer B.E. Johnson.

The figures for April 1944 give a random sample of the quantity of material recovered. During the month 91 serviceable and 1242 repairable items were recovered, 7805 lbs of aluminium alloy scrap, 7448 lbs of scrap metal and 550 lbs of scrap brass passed to the Department of Munitions facility at Villawood.

2 Central Recovery Depot recorded a fatality, when Leading Aircraftman T.A. McCombie was killed whilst on duty in a motor cycle

accident at Windsor on 16 September 1944. He was buried, with full honours, at Richmond on the 20th.

One of the conversion projects was that of the last Westland Wapiti in service, which commenced on 15 August 1944.

2 Central Recovery Depot was almost engulfed in a fire which broke out in the 2 Aircraft Depot 'K' Group storage area on 1 August 1945.

The strength of the unit in September 1945 was 278; in January 1946, it had dropped to 215 and, when the last entry was made in the Unit History Record on 30 June 1946, down to 31.

Commanding Officers

15 January 1944	—Flight Lieutenant W. Lewis
7 March 1945	—Squadron Leader C.S. Scott
19 November 1945	—Squadron Leader W.J. Taylor
1 March 1946	—Flight Lieutenant R.B. Cotterill
1 April 1946	—Flight Lieutenant W. Richardson

3 CENTRAL RECOVERY DEPOT

CHRONOLOGY

11 January 1944	—formed at Amberley, Qld
26 August 1946	—last entry in Unit History Record

NARRATIVE

On 11 January 1944, Flight Lieutenant J.E. Jackson formed 3 Central Recovery Depot from the technical salvage sections of 3 Aircraft Depot, 8 Service Flying Training School and the non-technical recovery and disposal section of 3 Stores Depot. Detachment 'A' was raised at New Farm, and moved to Breakfast Creek, Brisbane, on 10 December 1945. Another detachment operated from Oakey, Queensland, until 17 December 1945.

The first aircraft received by the unit was a Fairey Battle from 1 Air Observers School on 3 March 1944; the unit recovered its first aircraft (a 4 Communications Unit Anson) from the photographic aerodrome at Toowoomba.

Members of the unit travelled to various localities which included Maryborough, Charleville and Kingaroy to salvage Spitfire, Anson, Gipsy Moth and other aircraft types. Some salvage operations were difficult. On 27 August 1944, Pilot Officer H.A. Sweetser and 11 other ranks arrived from Joskleigh after recovering an Anson from the nearby beach. Consistent rain had made the coast road impassable, so the aircraft had to be transported by the inland defence road through Eidsvold and Nurrindi. Before reaching the road, the party took three hours to cross a 400-yard salt pan as the vehicles became bogged.

During December, the unit took over Hangar 256 from 3 Aircraft Depot, to give extra space for its increased activity. The following figures give some idea of the importance of the salvage operations: 48 747 lbs of steel, 418 242 lbs of mixed scrap metal, 4240 lbs of brass, 840 lbs of lead, and 162 888 lbs of aluminium. A total of 155 107 serviceable and 322 185 repairable items were recovered during the period of Depot operations.

Stores from the Depot were auctioned between 17 and 18 September 1946, and the last record in the Unit History Record is dated 26 August 1946.

Commanding Officers

11 January 1944	—Flight Lieutenant J.E. Jackson
24 May 1944	—Squadron Leader N.M. Dyer
11 February 1946	—Flying Officer W.M. Campbell
March 1946	—Flight Lieutenant E.E. Walden

4 CENTRAL RECOVERY DEPOT

CHRONOLOGY

15 March 1944	—formed at Boulder, WA
21 March 1944	—advance party to Maylands, WA
12 April 1946	—unit disbanded

NARRATIVE

The manpower resources of the salvage sections of 4 Aircraft Depot, Boulder, 4 Service Flying Training School, Geraldton, 9 Elementary Flying Training School, Cunderdin, and 4 Supply Depot, Maylands, were combined to form 4 Central Recovery Depot on 15 March 1944. In addition to the main body at Boulder, Flight Lieutenant G.H. Tournay, the Commanding Officer, had responsibility for two detachments—'A' at Geraldton and 'B' at Maylands. The latter was the non-technical section.

During 16 March 1944, Tournay travelled to Perth to inspect a site owned by the Kauri Timber Company at Maylands, which was to be the site of the unit. Sergeant Carstairs led an advance party of six airmen to the site on 21 March, to prepare it for the transfer of stores and equipment during April. Part of the preparation entailed the timber flooring of one of the buildings and the erection of a barbed wire fence around the perimeter.

The first engines, Cheetah IXs, were accepted for conversion to components on 27 March, and the unit headquarters transferred from Boulder to Maylands on 12 April 1944. The move to Maylands was complete with the arrival of the last nine airmen from Detachment 'A' on 4 May 1944.

A Tiger Moth, the first aircraft allocated to the Depot, was forwarded from MacRobertson and Millers for conversion to an instructional airframe on 28 April 1944. Next day, Flying Officer Windover and a party of six proceeded to Yanchep to salvage a Vultee Vengeance aircraft. During the period of Depot activity, the salvage crews recovered aircraft from Geraldton, Broome, Cunderdin and Learmonth.

The Depot expanded from a strength of five RAAF officers and one WAAAF officer, 127 other ranks and 34 WAAAF, to peak at 171 officers and other ranks (including 30 airwomen) by the end of February 1945. The facilities also expanded as, on 19 May 1944, Building 2 was dismantled and re-erected on the north-east boundary to provide for a guard room, dope shop, transport section and change room.

Members of the unit participated in organised sport, playing in the local competition in tennis, football and golf. During July 1944, a football team from the Depot played one made up of members of the Barton Mill Prison Camp.

From September 1945 a feature of Depot activities was its involvement with the disbandment of radar units. On 25 September 1945, Sergeant Nockels and party proceeded to Cape Naturalist to disband 33 Radar Station, and Flight Lieutenant Turner assisted with the disbandment of 48 Radar Station at Jurien Bay. Other radar stations which were dismantled by members of the Depot were 45 and 47, and Sergeant Vaux travelled to Busselton on 18 October 1945 to assist with the disbandment of 4 Recruit Depot.

It had been planned to move the Depot on two occasions. On 14 August 1944, a site at South Guildford was inspected, but found to be unsuitable. On 20 December 1944, discussions took place at Pearce regarding the move of the Depot to that site early in 1946. It was planned that stores would be moved into a 14 Squadron hangar, and that RAAF Pearce would take over control of these stores on a 'caretaker' basis pending the final authorisation for the move to 4 Central Recovery Depot. However, the Depot did not move to Pearce, and was disbanded on 12 April 1946.

Commanding Officers

15 March 1944	—Flight Lieutenant G.H. Tournay
24 May 1945	—Squadron Leader R. Darlington
5 November 1945	—Squadron Leader P. Neale

5 CENTRAL RECOVERY DEPOT

CHRONOLOGY

1 February 1944	—formed at Mallala, SA
17 September 1945	—moved to Port Pirie, SA
15 March 1946	—re-named Care and Maintenance Unit, Port Pirie
4 August 1947	—command reverted to 14 Stores Unit

NARRATIVE

5 Central Recovery Depot was formed at Mallala, South Australia, by Squadron Leader T.M. Carroll, with the strength of 28 airmen and three WAAAF. In addition, 33 airmen and one WAAAF were posted to 5 Central Recovery Depot and attached to 2 Operational Training Unit at Mildura. The Depot operated detachments at Waymouth Street, Adelaide, (Detachment 'A') and at Parafield.

Mallala was only a temporary location. The Finsburg Munitions Annex area was inspected on 16 February 1944 but, although the floor space was adequate, the building was found incapable of carrying overhead lifting gear. On 29 March, discussions were held regarding a possible move to Gawler, but this did not eventuate. Parafield was also inspected, but found unsuitable for Depot operations.

On 6 September 1944, the Depot was instructed to move to Port Pirie, South Australia. On 11 September, an advance party under the command of Flying Officer Johnson proceeded to Port Pirie, and the move was completed (except for a three-man rear guard who remained at Mallala to service Battle aircraft) on 17 September.

Whilst the final locality of the Depot was being discussed, salvage tasks commenced. On 17 March 1944, Pilot Officer Johnson proceeded to Ceduna to assess the damage to a Kittyhawk. Members of the unit ranged to Cooks Plains, Mount Bryan, Ceduna, Yunta, Murray Bridge, Broken Hill and Alice Springs to recover Battle, Vengeance, Kittyhawk, Spitfire, Beaufort, Tiger Moth and Hudson aircraft.

At the end of 1944, the strength of the unit was 142 officers and other ranks, and the unit reached a peak strength of 228 in January 1946.

During September 1945, plans were made for members of the Depot to assist with the closure of inland fuel depots at Port Pirie, Crystal Brook and Gladstone. Additional tasks taken over by the unit were the storage of aircraft—during November/December Anson aircraft arrived—and the responsibility for the RAAF station at Port Pirie on the closure of 3 Air Observation School on 18 December 1945.

5 Central Recovery Depot ceased to function on 10 January 1946, and was to be reduced to a nucleus basis. When it was re-named Care and Maintenance Unit, Port Pirie, on 15 March 1946, the strength of the unit had dropped to 95 officers and other ranks. Strength dropped to four during September, which hampered the listing and segregating of equipment for Commonwealth Disposal Commission auctions. The first auctions took place on 2–3 December 1946, with all purchases, including five Tiger Moth aircraft to the Royal Aero Club of South Australia, being delivered three days later.

During February 1947 three airmen were attached from 14 Stores Unit to assist with the listing and segregation of equipment. In April 1947 the strength of the unit had fallen to two.

In June 1947, Anson aircraft were delivered to various customers, and representatives of the Public Stores Department and Department of Works selected equipment from the 5 Central Recovery Depot stock for their departments' use during July.

On 4 August 1947, Squadron Leader Cottee was posted to 1 Aircraft Depot and the control of the unit at Port Pirie was assumed by Squadron Leader Power of 14 Stores Unit.

Commanding Officers

1 February 1944	—Squadron Leader T.M. Carroll
10 May 1945	—Squadron Leader A.H. Cottee

6 CENTRAL RECOVERY DEPOT

CHRONOLOGY

15 June 1944	—formed at Breddan, Qld
June 1945	—moved to Stuart, Qld
April 1946	—last entry in Unit History Record

NARRATIVE

Personnel and equipment from the Technical Salvage Section of 13 Aircraft Repair Depot became the basis for 6 Central Recovery Depot, which was raised at Breddan on 15 June 1944, under the command of Squadron Leader A.H. Barter.

On 17 June a salvage party was operating at Proserpine, recovering a Spitfire. Parties travelled to Ravenswood, Alligator Creek, Longreach, Cooktown, Rockhampton, Mackay, Pipan Island and Burdekin River during its period of operations. Not all recovery activity went as planned. On 13 November, the transport of Boomerangs along Range Road to Mareeba was found difficult because the road was narrow and curving; the latter was a factor when one of the aircraft was damaged when it was hit by an Army truck.

6 Central Recovery Depot was involved in an unusual recovery, which commenced on 28 November 1944. A Sunderland flying boat had been holed in the bow at Townsville, and sank on a sand bar. The aircraft was covered by the evening tide, and the equipment and flotation bags supplied by the Depot were used to raise the aircraft during the following morning and beach it near Queens Hotel, Townsville, next day.

A double fatality occurred on 4 November 1944. A Wirraway aircraft struck and damaged the tail unit of a Vengeance aircraft before crashing on the opposite side of the airstrip. There were five men working on the rear section of the Vengeance, and two were killed as a result of the accident.

Although the Depot expanded at Breddan, there was confusion regarding its ultimate site. An igloo hangar was completed on 17 November, but this became flooded in the rainy season, and Army facilities at Breddan had to be used as the unit receipt section during January 1945. During November 1944, proposals had been made to move 6 Central Recovery Depot to the site of the United States Army Air Force

4th Aircraft Depot Group at Garbutt. On 10 March 1945, North-Eastern Area authorised the move of the Depot to a hangar adjacent to the Stuart Highway Railway Station, with a camp site to be located at Roseneath. The move, delayed by the flooding of the Burdekin River, commenced eight days later, but orders were received on 27 March to cease the move immediately. Although 40 per cent of the equipment had been moved, 6 Central Recovery Depot was ordered back to its old location at Breddan, where hangars had to be reconstructed. At the same time steel huts were built to house the Engine Repair Section. On 29 May 1945, Warrant Officer Messell and 15 men proceeded to Roseneath camp to prepare it for habitation, and the Depot moved to the Stuart site during June 1945.

At the end of July 1944, the personnel strength of the Depot totalled 41. This peaked at 212 during September 1945. From this date there was a steady decline in strength. When the final entry was made in the Unit History Record in April 1946, the strength of the unit was 41.

Commanding Officer

15 June 1944 —Squadron Leader A.H. Barter

7 CENTRAL RECOVERY DEPOT

CHRONOLOGY

15 June 1944	—raised at Tocumwal, NSW
7 October 1946	—last entry in Unit History Record

NARRATIVE

Flight Lieutenant A.J. Kibble was the temporary Commanding Officer of 7 Central Recovery Depot when it was raised at Tocumwal, New South Wales, on 15 June 1944. The unit occupied the 7 Aircraft Depot Technical Salvage Section hangar and took over the function of that section.

Squadron Leader H.H. Powell assumed command of the Depot on 6 July 1944, to find that the activities of the unit were restricted due to having to undertake a physical stocktake of items transferred from 7 Aircraft Depot, and difficulty in obtaining hangar space and equipment. The posting of Flight Lieutenant J.A. Williams to the Depot as engineering officer on 1 September ensured that the best results were obtained from the resources available, and the equipment position slowly improved. In September 1944, lathe drilling machines and wood working machines were received, and the unit was considered to be operational in November. During December, the Depot strength reached a peak of 119 personnel.

Although the wood and machine shops were not wired until January 1945, six engines and four aircraft were converted to scrap and spare parts during February. This activity was restricted by the lack of technical personnel. Despite this, between April and September 1945 approximately 54 aircraft and 20 engines were converted to scrap and spare parts; aircraft types included Wackett Trainer, Anson, Liberator, Boomerang and Oxford.

In February 1946 Flight Lieutenant J.B. Moodie was the temporary Commanding Officer, with 54 personnel under command. When Flight Lieutenant J. Harker assumed command on 11 March 1946, the unit strength was nine. This dropped to three when the Commanding Officer signed his final report on 7 October 1946.

Commanding Officers

6 July 1944	—Squadron Leader H.H. Powell
9 January 1945	—Squadron Leader G.F. Gardick
2 August 1945	—Squadron Leader J. Tennant
11 March 1946	—Flight Lieutenant J. Harker

8 CENTRAL RECOVERY DEPOT

CHRONOLOGY

1 June 1944	—formed at Darwin, NT
11 August 1944	—moved to 11 Works Supply Unit site
4 December 1945	—Depot disbanded

NARRATIVE

8 Central Recovery Depot was established at Darwin on 1 June 1944, under the command of Flight Lieutenant R.G. Watt. A detachment was formed at 14 Aircraft Repair Depot, Gorrie, to handle salvage taken over from that unit, and officers and sergeants were accommodated at a house named 'The Savoy'. The airmen were housed in 'Gorrie House', and in a comet hut.

On 3 June 1944, the administrative office and hangar at the Darwin Civil Airstrip were converted to offices. The unit established an equipment store in a comet hut erected for the purpose, and two 80-foot by 18-foot bush structures were erected as the airmen's mess and armament and electrical sections. The mess was completed on 5 July.

Twelve Spitfire, two Boomerang, three Battle, two Beaufighter, one Walrus and a Beaufort aircraft arrived during June for disposal or conversion to components.

On 8 June 1944, orders were received from Headquarters North-Western Area to move the Depot to the site vacated by 11 Works Supply Unit. The move commenced on 11 August 1944.

The unit erected facilities at the new site, and continued with salvage operations, which included the recovery of a Liberator from Truscott, a Mosquito from Coomalie Creek, and a Catalina. During the same period, members of the unit participated in unit sports, including rugby league, Australian football, cricket, tennis, and basketball.

The unit was disbanded on 4 December 1945.

COMMANDING OFFICERS

1 June 1944	—Flight Lieutenant R.G. Watt
1 September 1944	—Squadron Leader N.E. Cottee
1 June 1945	—Flight Lieutenant J.C. Kane

1 FLYING BOAT MAINTENANCE UNIT

CHRONOLOGY

10 October 1943	—formed at Bowen, Qld
8 January 1947	—last entry in Unit History Record

NARRATIVE

1 Flying Boat Maintenance Unit was formed at Bowen, Queensland, on 10 October 1943, from personnel posted from 22 Operational Base Unit. The strength of the Unit on formation was 12 officers and 284 airmen with Catalina aircraft A24-49, A24-57, A24-65, A24-64, A24-54 and A24-68. The first Commanding Officer, Squadron Leader O'Donnell, arrived on 4 November 1943.

On 16 June 1944, the Minister for Air and his party of officials arrived by air to inspect the Unit. In addition, an investigation into the hiring of barracks at the Unit took place.

On 31 July 1944 the Civil Construction Corps started work on the waterfront for the enlargement of the concrete tarmac. During August considerable progress was made on the extensive works and building program in the slipway area. On 2 January 1945 Unit headquarters moved from Herbert Street to the new administration building in the slipway area.

The Unit had its share of accidents, and was involved in other incidents and searches. On 19 September 1944, a United States Army Air Force Catalina, allotted to the Unit for repairs, had seriously damaged planing surfaces, and landed on a nearby strip. The mainplane outer panels were removed and transported from the landing strip to the Unit's slipway area. The aircraft was transported by road—an intricate undertaking owing to the aircraft having to negotiate a number of railway crossings and the main streets of Bowen.

On 12 October 1944 a privately owned launch in difficulties off Bowen was brought in by one of the Unit's marine craft and on 19 December 1944 Catalina A24-101 searched for a Norseman aircraft which had made an emergency landing en route from Rockhampton to Mackay. However, the Norseman was located by another search aircraft.

On 21 December 1944, a search was made, with negative results, for the naval launch HMAS *Lealani*, which had been damaged in a collision en route from Rockhampton to Townsville. The good news was received that HMAS *Lealani* had arrived at Port Denison.

On 19 January 1945 Catalina A24-92, on a ferry flight from 2 Flying Boat Repair Depot, Rathmines, New South Wales, to 111 Air–Sea Rescue Flight, was damaged when landing at Bowen. Flight Lieutenant W.H. Cox of 20 Squadron was injured, his left forearm being severed by a propeller.

On 17 May 1945 a fire was reported in a civilian garage. The Unit fire tender assisted in confining what could have been a serious conflagration.

Mariner aircraft A70-7 arrived on 6 August 1945 to inaugurate the Air–Sea Rescue base at Bowen for 114 Air–Sea Rescue Flight.

On 2 September 1945 vacation of civilian properties commenced. All WAAAF personnel were posted from the Unit on 29 October 1945, the first contingent departed for Townsville on 30 October 1945 and the balance by the end of that week. The Unit History Record acknowledged the good work of the WAAAF members of the Unit. With the departure of WAAAF personnel, the Hotel Denison, which had been used for their accommodation, was returned to the civilian owner on 13 November 1945.

On 8 January 1947 one of the two main hangars on the slipway 'building Number 7', was vacated by the RAAF and taken over by the Department of Civil Aviation.

Commanding Officers

4 November 1943	—Squadron Leader J.A. O'Donnell
2 June 1944	—Squadron Leader E.F. Easterbrook
14 December 1944	—Squadron Leader G.L. Grendon
2 February 1945	—Squadron Leader E.J. Stocker
25 February 1946	—Flying Officer R.C. Henzell

2 FLYING BOAT MAINTENANCE UNIT

CHRONOLOGY

20 July 1944	—formed at Lake Boga, Vic.
25 September 1944	—commenced move to Darwin, NT
7 October 1945	—last entry in Unit History Record

NARRATIVE

2 Flying Boat Maintenance Unit was formed at Lake Boga, Victoria, as a lodger unit to 1 Flying Boat Repair Depot, on 20 July 1944, under the command of Squadron Leader J.A. O'Donnell, with a strength of 44 other ranks.

On 2 August 1944 personnel commenced pre-embarkation leave.

Work commenced on dismantling a Netherlands East Indies B-25 aircraft (N5-182) on 12 August 1944.

Personnel were tropically kitted; a lecture was given on tropical diseases, drinking water and VD before movement north.

On 25 September 1944 the main party of four officers and 53 other ranks departed by rail, en route for Darwin. On 10 October work commenced on the establishment of a temporary orderly room and marine section at Doctors Gully. Over the next few months work continued in establishing Doctors Gully which included clearing of land, the erection of accommodation huts, workshops, a temporary slipway, heavy earthworks and other related works.

On 24 November 1944 a directive was received for the Unit to move to East Arm as soon as Unit personnel had erected buildings. Consequently 12 personnel were detailed to commence work at East Arm to clear land for prefabricated hut accommodation. Work at this site was made difficult by the wet season, insects and inadequate equipment. While work continued slowly on the East Arm area, many of the Unit personnel were detached to other areas such as 55 Operational Base Unit, Gorrie, Winnellie and RAAF Darwin.

On 11 March 1945, Unit headquarters was officially established at East Arm. By the end of the month, the move from Doctors Gully to East Arm was completed.

Catalina A24-99, which had been allocated to the Unit some months before, had repairs completed by the end of April 1945, and proceeded to 1 Flying Boat Maintenance Unit at Bowen. The repairs to aircraft gained momentum, though affected by continuing building work and the short supply of aircraft spare parts.

On 10 August 1945, news was received that Japan was prepared to accept the terms laid down at the Potsdam Conference. It was not long before all messes were alive, and a merry night was enjoyed by all. Darwin Harbour was alight with fireworks with a display of pyrotechnics by the Navy and local sources.

On 15 August 1945 a broadcast reported that Japan had surrendered unconditionally, and orders were given to stand down. A thanksgiving service was held followed by pay parade and the issue of two bottles of beer to personnel. Parties were held in the huts and the celebrations were continued well into the night.

Aircraft repairs still continued, and the tarmac was in the process of being extended. The last entry in the Unit History Record was made on 7 October 1945.

Commanding Officers

14 July 1944	—Squadron Leader J.A. O'Donnell
7 March 1945	—Squadron Leader G.L. Grendon

1 FLYING BOAT REPAIR DEPOT

CHRONOLOGY

16 June 1942	—formed temporarily at RAAF Station Rathmines, NSW
2 July 1942	—advance party arrived at Swan Hill, Vic.
20 July 1944	—2 Flying Boat Maintenance Unit formed as lodger unit
1 March 1946	—declared Care and Maintenance Unit
12 November 1947	—Care and Maintenance Unit, Lake Boga, disbanded

NARRATIVE

1 Flying Boat Repair Depot was formed at RAAF Station Rathmines, New South Wales, on 16 June 1942 as a separate Air Force unit within the command of Eastern Area. The Rathmines, New South Wales, establishment was a temporary expedient pending completion of the depot and quarters at Lake Boga, Victoria. Temporary command of the unit was assumed by Flight Lieutenant G.S. Moffatt on 18 June 1942.

In accordance with the RAAF expansion program a suitable area was urgently required for a flying boat depot with landing facilities for both land-based and waterborne aircraft. These requirements were fulfilled at Lake Boga, Victoria, where a site was selected, approximately 16 kilometres south-east of Swan Hill on the north-west foreshore of Lake Boga adjacent to the Murray Valley Highway and the Swan Hill–Melbourne railway. Electric power was available from the Swan Hill power station.

While the requisite buildings were constructed and equipped, the upgrading of the Swan Hill aerodrome proceeded in preparation for arrival of land-based aircraft.

The camp complex some three kilometres distant from the base was designed and constructed to appear as a natural extension of the Lake Boga township, for purposes of camouflage.

On 2 July 1942 an advance party comprising 30 personnel of 1 Repair Depot arrived at Swan Hill and were billeted at the Royal Hotel. A barracks store was leased at 83 Campbell Street, Swan Hill, on 4 July

1942, with temporary Depot headquarters being established at 91 Campbell Street on 6 July.

The Depot was declared a master stores depot for Catalina equipment on 22 July. The first Catalina to be mechanically overhauled alighted on 5 August 1942 while the initial permanent slipway was completed by 6 October 1942.

Movement of 150 personnel from the Royal Hotel, Swan Hill, to the Depot site at Lake Boga occurred on 14 October 1942.

Squadron Leader G.D. Marshall assumed command of 1 Flying Boat Repair Depot on 16 November 1942. Depot strength at 31 December 1942 stood at 618 personnel.

March 1943 proved a testing time when fire broke out in the Depot sick quarters. On the evening of 21 March Depot headquarters was notified that the sergeants' mess was aflame; the building was gutted in approximately 30 minutes.

By mid-1943 1 Flying Boat Repair Depot was carrying its full share of the Allied wartime burden with its personnel executing prompt turnaround of maintenance aircraft for the RAAF, United States Navy and the Netherlands East Indies Naval Air Service.

2 Flying Boat Maintenance Unit was formed in nucleus as a lodger unit of 1 Flying Boat Repair Depot on 20 July 1944 and after eight weeks training departed for Darwin, North-Western Area.

The month of October heralded a period of extremely uncomfortable weather conditions for the district. Boisterous to gale force winds accompanied some of the worst dust and sand storms ever encountered in Lake Boga's history. At times the ongoing dust storms caused power disruption from the Swan Hill power station and plunged the camp into darkness while the Depot emergency power unit was activated to service the base area.

Aircraft maintenance proceeded, with work also being executed for other units including the modification of Liberator and Mosquito parts and the manufacture of Kittyhawk rudders.

On 18 December 1944, 635 kilograms of dust was swept from a Catalina and its hangar floor after a dust storm. The dust storms continued and many personnel were issued with goggles.

In December 1944, the Depot reached its peak strength of 943 personnel comprising 39 officers, 802 airmen and 102 WAAAF.

As from 1 March 1946, 1 Flying Boat Repair Depot was re-named Care and Maintenance Unit. Later that month the unit's sick quarters were closed. By the end of March most vehicles had departed to Melbourne. The unit post office closed down on 1 May and on 3 May 1946 authority was granted to close all signals facilities.

3 October 1946 saw Squadron Leader E.J. Symons assume command from Squadron Leader J.A. O'Donnell to supervise the closing stages of the unit. At that date there were 59 flying boats and 13 sea planes stored at the unit.

On 12 November 1947 the Care and Maintenance Unit was disbanded.

Commanding Officers

16 November 1942	—Squadron Leader G.D. Marshall
29 September 1943	—Wing Commander G.U. Allan
11 December 1944	—Wing Commander R.S. Rice
7 January 1944	—Wing Commander R.H. Foord
13 November 1945	—Squadron Leader C.R. De La Rue
25 April 1946	—Squadron Leader J.A. O'Donnell
3 October 1946	—Squadron Leader E.J. Symons

2 FLYING BOAT REPAIR DEPOT

CHRONOLOGY

10 February 1943	—formed at Rathmines, NSW
26 April 1946	—Depot disbanded

NARRATIVE

2 Flying Boat Repair Depot was formed at RAAF Station Rathmines, New South Wales, on 10 February 1943.

The Depot carried out repairs, overhauls and modifications to flying boats. Catalina aircraft had camera installation, armament installation and armament modifications carried out by the Depot, and aircraft were camouflaged.

The Depot serviced a large variety of aircraft, including Catalina, Dolphin, Dornier, Seagull and Walrus, from the RAAF, United States Navy and the Royal Navy.

The Depot was involved in salvage operations of wrecked aircraft. The first occurred in February 1943 when a Wirraway aircraft crashed in Lake Munmorah. Another Wirraway crashed near Lake Munmorah at the same time and the Commanding Officer and Flying Officer Burrow went to investigate. On 24 May 1943, Catalina A24-39 crashed into Port Stephens, killing seven personnel, and was later salvaged by the Depot.

On 15 August 1945, with the cessation of hostilities in the Pacific, Depot personnel were paraded and addressed by the Commanding Officer, followed by special leave. On 27 November 1945, Mr C.C. Altmann from the Ministry of Post War Reconstruction delivered a lecture on post-war problems and rehabilitation to the Depot.

The Depot disbanded on 26 April 1946.

COMMANDING OFFICERS

10 February 1943	—Squadron Leader E.F. Easterbrook
8 March 1943	—Wing Commander G. Allan
18 May 1944	—Wing Commander R.S. Rice
19 December 1944	—Squadron Leader E.F. Easterbrook
29 March 1946	—Squadron Leader J. Tennant

478 MAINTENANCE SQUADRON

Jaga Mutu Pedang (Guard the Sword)

CHRONOLOGY

27 June 1947	—formed at Williamtown, NSW
1 April 1948	—disbanded
24 January 1949	—reformed
28 July 1952	—moved to Malta
March 1955	—moved to Williamtown
November 1958	—moved to Butterworth, Malaya
31 October 1983	—Squadron disbanded

NARRATIVE

Of the five main maintenance squadrons formed to support units flying particular types of aircraft, 478 Maintenance Squadron could be classified as the 'glamour' squadron, if a maintenance squadron could be so classified. In its 36 years supporting fighter squadrons, it was responsible for servicing RAAF Wirraway, Mustang, Vampire, Meteor, Sabre and Mirage aircraft, and RAF Vampires and Meteors, plus a variety of other service and civil aircraft, as necessary.

Formed on 27 June 1947 at Williamtown, 478 Maintenance Squadron was responsible for the control of all aircraft and equipment maintenance under the command of 78 (Fighter) Wing and within the command of Headquarters Eastern Area. Initially seven Wirraway aircraft were ferried from RAAF Tocumwal to Williamtown to be made fully serviceable for the training of replacement pilots for service with the British Commonwealth Occupation Force in Japan. Training commenced on 3 October 1947. One month later Mustang aircraft were made available for pilot training. Five months later, on 1 April 1948, 478 Squadron was disbanded.

On 24 January 1949 the Squadron was reformed and tooled up to service jet aircraft. As well as Wirraways and Mustang, Vampire jet aircraft were included in the establishment of aircraft of 78 (Fighter) Wing. In September 1949, one of the Vampire aircraft, serviced by 478, created new flying records between Sydney and Melbourne and return. Piloted by Flight Lieutenant I.R. (Pip) Olorenshaw, the Vampire was flown from Sydney to Melbourne in 67 minutes and completed the return trip in 55 minutes.

On 28 July 1952, in response to an invitation from the British Government for the RAAF to operate with the NATO Defence Forces, Mediterranean Area, 78 (Fighter) Wing with 75 and 76 Squadrons and 478 Squadron moved into the Royal Naval Air Station, HMS *Falcon* at Hal-Far, Malta. RAF Vampire and Meteor aircraft and servicing equipment was left at Malta as the RAF moved out and the RAAF assumed the responsibility for the fighter aircraft defence of the island. Later the RAAF moved to its own base at Takali Airstrip, Malta.

As the sole fighter aircraft defence force these squadrons soon became heavily involved in NATO defence exercises. Air raid exercises attacking Malta, averaging at least three each month, came from the United States Sixth Fleet, the United States Air Force in Libya, Royal Air Force Bomber Command, the Royal Navy, the French Air Force in North Africa, the Italian Air Force and elements of the Greek and Turkish Air Forces. Frequently the exercises were from dawn until dusk with the RAAF pilots flying up to five one-hour sorties each per day, and the ground staff servicing, refuelling and re-arming aircraft for each sortie. The pilots and ground crews soon earned high commendation from the attacking force commanders for the ability displayed in their respective roles.

Between the defence exercises the RAAF was to serve in other areas. A detachment of 16 Vampires and 95 ground staff moved to Horsham St Faith in England, in May 1953, to take part in the Coronation Air Review by Her Majesty Queen Elizabeth II. On 15 July the Vampires were among 649 airborne aircraft in the most comprehensive formation flypast ever planned, when Her Majesty reviewed the RAF. On 20 July the RAAF detachment was transferred from Horsham St Faith to RAF Station Wahn, in Germany, where, attached to 122 Wing RAF, it took part in NATO Exercise 'Coronet' with units from England, France, United States of America, Belgium, Portugal, Turkey and Greece. About 2000 aircraft and 40 000 troops were involved in the exercise. Before returning to Australia in 1955 detachments of aircraft and ground staff participated in NATO exercises at Idris and at Nicosia, Cyprus.

Flying at Malta ceased on 30 November 1954; the Wing left Malta during January 1955 and in March 1955 478 Squadron was re-established and commenced to function at Williamtown servicing Vampire, Meteor and the new Sabre fighter aircraft. Three years later 478 was to move again, this time to Butterworth, Malaya, with 78 (Fighter) Wing, 3 Squadron and 77 Squadron, as the RAAF component of the British Commonwealth Far East Strategic Reserve.

478 Squadron was involved again in a history-making exercise when Group Captain G.A. (Glen) Cooper, the Officer Commanding 78 Wing, led a flight of four 3 Squadron Sabres from Williamtown to RAAF Butterworth on 2 November 1958, having staged through Townsville, Darwin, Biak, Guriuab (Philippines) and Labuan (Borneo), thus completing phase one of Operation 'Sabre Ferry'. By February 1959 27 Sabres had been ferried through the staging posts without incident. 478 Squadron played a major role in the exercise, having serviced the aircraft for the long flight and the handling equipment required at each staging post to prepare the aircraft for the next leg of the flight.

Similar exercises became a feature of aircraft operations at Butterworth. Fighter defence exercises with British, American and SEATO forces required the Sabres to be flown to various parts of South-East Asia and the Philippines. Long hazardous flights over the South China Sea by the heavily laden single-engine aircraft from Butterworth to Saigon, thence to the United States Air Force Base at Clark Field, Manila, and return, were accomplished successfully due to the efficiency of the fighter pilots and the maintenance crews who serviced the aircraft.

In 1965 Mirage aircraft were brought into service. 75 Squadron was equipped with Mirages at Williamtown, then transferred to Butterworth to relieve 3 Squadron and its Sabres. 478 Squadron had to retool to service the Mirages. Later 77 Squadron was equipped with Mirages. In November 1967, 78 Wing was disbanded—478, 75 and 77 Squadrons became independent non-self-accounting units under the command of Headquarters RAAF Butterworth.

478 Squadron celebrated its 21st year since formation on 27 June 1968. The Squadron had pride of place on a special parade at Butterworth to celebrate the occasion. It was reviewed by the Officer Commanding RAAF Butterworth, Air Commodore G.H. Steege, and presented with

mementos of recognition for having completed an illustrious period of service, of which more than 12 years had been served overseas.

Serving in areas foreign to Australian standards, the Squadron became proficient in adapting to meet situations. Officers and ground staff distinguished themselves in technical and scientific fault investigations and developing and improving technical, ground handling and servicing equipment skills. The proficiency and attention to detail were a credit to all ranks. No job was too small or too large. In 1980 the Squadron housed and serviced motorised gliders being ferried from Romania to Tocumwal. Later it completely refurbished Dakota A65-69 before it was flown to Berlin to become a museum piece commemorating the RAAF's participation in the Berlin Air Lift after World War II.

In May 1983 the Squadron celebrated 25 years of service at Butterworth with functions and parades. On 31 October 1983 the Squadron was disbanded.

Commanding Officers

1947	—Squadron Leader J.C. Kane
1949	—Squadron Leader C.J. Leopold
1951	—Squadron Leader F.M. Timms
1952	—Squadron Leader J.C. Kane
1953	—Wing Commander E.K. De Cean
1956	—Wing Commander K.P. Connolly
1961	—Wing Commander J. Hughes
1963	—Wing Commander J.C. Kane
1965	—Wing Commander L.S. Compton
1967	—Wing Commander I.F. Sutherland
1969	—Wing Commander H.J. Bell
1971	—Group Captain F.A. Cousins
1972	—Group Captain C.K. Mahoney
1974	—Group Captain J.K. Henze
1976	—Group Captain C.W. Spitzkowski
1979	—Group Captain N.R. Wade
1981	—Group Captain I.K. Ashbrook

481 MAINTENANCE SQUADRON

Reliability

CHRONOLOGY

22 December 1945	—formed at Labuan, Borneo
23 March 1946	—moved to Japan
30 November 1948	—disbanded
3 January 1961	—reformed at Williamtown
31 January 1987	—Squadron disbanded

NARRATIVE

481 Maintenance Squadron was formed on 22 December 1945 at Labuan, Borneo, with its primary responsibilities for the maintenance and testing of new and old aircraft.

Air Vice Marshal C.A. Boucheir paid the Squadron a visit to brief the airmen about the conditions to expect in Japan, as the Squadron was to be re-located to Bofu. He addressed such topics as climatic conditions, various sporting possibilities, and the responsibilities of the Squadron as a whole. The move started on 23 March 1946, with the Squadron first transporting the equipment on two landing ships. The following day personnel boarded HMAS *Glengyle*, and started the slow journey to their new location, arriving in Kure on 1 June 1946. After unpacking supplies, Squadron personnel left for Bofu by rail.

The Squadron's enthusiastic sporting groups started almost immediately when everyone arrived at Bofu. The Aussie Rules team successfully played their first game against 381 Base Squadron on 2 June 1946. The Squadron's Rugby League team was also formed in preparation for a scratch match.

The Sergeant's Mess was open for the first time at Bofu on 10 July 1946. The Commanding Officer was invited to help celebrate the evening.

A small earthquake occurred on 21 December 1946 at 0415 hours. The quake was quite severe, but luckily no structural damage was sustained to Squadron buildings.

481 Maintenance Squadron expected to move to Iwakuni on 26 January 1948, but eventually left on 13 March 1948. The Squadron resumed operations at the new location on 15 March 1948. The Squadron had difficulty getting power to operate its electrical machinery and new transformers had to be erected to accommodate this problem.

The Squadron disbanded on 30 November 1948. Its stay in Japan was a short but productive one. The eagerness of Squadron personnel to perform any duty at hand showed dedication and commitment toward their service duties which was a credit to 481 Maintenance Squadron and the RAAF as a whole.

481 Maintenance Squadron was reformed on 3 January 1961 at RAAF Base Williamtown, just north of Newcastle, as part of 81 Wing. The main purpose of the newly formed 481 Squadron was to develop an efficient technical administration.

The Commanding Officer organised a static display of aircraft and equipment for 400 Legacy wards and widows when they visited the base on 5 September 1961. The Minister for Air, the Honourable David E. Fairbairn, also visited the base for a familiarisation tour on 26 September 1961.

RAAF Williamtown opened to the public on 14 September 1963. A static display showing the major activity of 481 Squadron was arranged in Hangar 173. Activities were organised by Squadron Leader R.H. Cowburn. The Minister for Air, the Honourable Peter Howson, was taken on a tour of 481 Squadron by the Commanding Officer Wing Commander J. Hughes and Squadron Leader J.K. Henze.

The Squadron crest, signed by Her Majesty Queen Elizabeth II, was received on 28 May 1970. Once the document was signed it was hung in a distinguished area of the Squadron. Prince Charles made an informal visit to the Squadron on 18 October 1974 before heading off to Sydney. Members of 481 Squadron acting as duty crew attended his VIP aircraft.

During 17–27 June 1975, Exercise 'Spanish Dollar' was conducted out of RAAF Base Williamtown. 481 Squadron was required to provide maintenance facilities from 4 Flight Hangar. 481 Squadron also participated in the 55th Anniversary Air Display and Open Day on 3 April 1976.

In early November 1976, a McDonnell Douglas promotion team visited the Squadron. Its purpose was to give a demonstration regarding the F-15 Eagle fighter aircraft.

Queen Elizabeth II toured Australia in 1977, and visited RAAF Williamtown on 11 March. During her stay, 481 Squadron provided the ground crew and part of the quarter crew to her aircraft.

481 Squadron disbanded on 31 January 1987.

COMMANDING OFFICERS

22 December 1945	—Squadron Leader S. Medley
23 January 1946	—Squadron Leader J. Hearnden
3 January 1961	—Wing Commander R.W.T. Ayre
11 March 1963	—Wing Commander J. Hughes
25 February 1966	—Wing Commander H.T. Bell
4 January 1967	—Group Captain D.R. Cuming
1 July 1968	—Wing Commander D.H. Smith
8 February 1971	—Group Captain H.T. Bell
17 January 1972	—Group Captain J.A. Dietz
26 January 1974	—Group Captain C.H. Mahoney
7 July 1975	—Group Captain F. Howie
23 January 1979	—Group Captain W.E. Sansum
13 January 1981	—Group Captain J.B. Macnaughton
7 December 1982	—Group Captain G.H.G. Homer
7 December 1983	—Group Captain J.T. Connell
November 1986	—Group Captain N.A. Smith

482 MAINTENANCE SQUADRON

Trenchant

CHRONOLOGY

10 May 1946	—4 Repair and Salvage Unit re-named 482 Maintenance Squadron
2 December 1991	—482 Squadron amalgamated with 3 Aircraft Depot to form 501 Wing
30 June 1992	—Squadron disbanded

NARRATIVE

On 10 May 1946, 4 Repair and Salvage Unit completed its move from Parkes to Amberley, and was re-named 482 Maintenance Squadron. The unit became a component of 82 Wing, responsible for the maintenance of the Liberator aircraft held on strength at 12, 21 and 23 Squadrons. From July to September 1947, when 82 Wing was in the process of converting from Liberator to Lincoln aircraft, a number of 482 Squadron personnel proceeded to East Sale to undergo Lincoln familiarisation courses.

The shortage of Merlin engines reached a critical stage in May 1950, creating a huge maintenance problem. Activity increased during the following year, however, when packing of spares, tools and handling equipment for Operation 'Buffalo Bill' began on 23 May. This was followed on 1 October by the commencement of the first Avon engine course, and the first Canberra airframe course on 3 October. By 22 October, preparations were under way for Operation 'Pacific Outpost'.

In February 1954, the Squadron was kept busy servicing A65-85 of the Queen's Flight, and taking part in Operation 'Royal Salute'. On 18 February, 3000 lbs of food and blood plasma were dropped over the flooded towns of Springsure and Rolleston in central Queensland, followed by further drops of flour and yeast on 21 February. The following day, rescue apparatus, life jackets, food and blankets were dropped by Dakota to people cut off by floodwaters in the Lismore district. On 2 March, 13 storepedoes, each packed with 200 lbs of food, were loaded for a supply drop on Fairfield Station within the flooded area of central Queensland.

From 1 to 3 June 1955, the Squadron provided marshalling, parking and refuelling facilities for elements of the United States Air Force visiting Amberley. The visiting aircraft consisted of two Globemasters, two Superfortress tankers, four Thunderjets and one Fortress. Provisions were also made for a public inspection of these aircraft. In June 1956, the Squadron provided ground support for 82 Wing Canberra aircraft engaged on Operation 'Bala Lagin', and in February 1958, participated in a royal guard of honour at a civic reception for the Queen Mother at her departure from Queensland. One officer and 100 airmen from the Squadron also took part in the combined services Queen's Birthday review held at Victoria Park, Brisbane, on 15 June 1959.

Members of the Squadron's safety equipment section were included in a RAAF display at the aquatic carnival at Milton Reach, Brisbane River, for Australia Day celebrations on 1 February 1960. A guard of honour for the opening of State Parliament on 24 August 1960 included 40 airmen from the Squadron.

The Commanding Officer and a party of airmen supported a 1 Squadron exercise in New Guinea in November 1960, and were deployed to Darwin the following month to support 82 Wing Exercise 'High Noon'.

One working weekend a month was undertaken by the Squadron during February–March 1961 to provide training for the Citizen Air Force members of 24 Squadron. In August–September 1962 35 airmen formed part of a guard of honour on the arrival and departure of the King and Queen of Thailand. On 20 January 1967 a maintenance party from the Squadron proceeded to Eagle Farm Airport to attend to the marshalling and maintenance of aircraft engaged in the transport of the Prime Minister of South Vietnam.

When responsibility for the maintenance of all Canberra aircraft was transferred to 1 Operational Conversion Unit (1OCU) at the end of April 1968, the electrical, instrument, maintenance control, maintenance hangar and radio sections of 482 Squadron, including personnel, were transferred to 1 Operational Conversion Unit.

The first ferry flight of five Phantoms arrived on 14 September 1970, followed by the second flight of six Phantoms five days later. Eight Phantom aircraft plus a maintenance party departed on 6 April 1972 for air displays at Fairbairn and Laverton, followed on 5 May by the departure of 13 Phantoms and 97 personnel, forming 482 Squadron Detachment 'A', for Exercise 'Top Limit' in Darwin.

On 15 February 1977, six F-111C aircraft and 74 Squadron personnel were deployed to Bucholz Army Airfield at Kwajalein in the Marshall Islands, thence to Hickam Air Force Base, Hawaii, for the joint maritime exercise 'Rimpac 77'. 482 Squadron personnel were deployed to Guyra, New South Wales, on 28 April 1977 for guard duties and to assist in recovery and investigation at the crash site of F-111C A8-136, from which the crew had successfully ejected. On 14 May 1980, a party from the Squadron departed to RAAF East Sale for the rescue and recovery of F-111C A8-128.

During 1981, the Squadron was visited by many groups from preschools, primary and high schools, and various other youth and community groups. On 1 February, at the direction of the Chief of the Air Staff, responsibility for operating level maintenance of F-111 aircraft was transferred from 482 Squadron to 1 and 6 Squadrons. All F-111 aircraft, and a number of technical facilities plus 202 technical and other staff, were also transferred to 1 and 6 Squadrons.

Work continued at the Squadron on F-111 servicing and modifications. F-111 and non-F-111 courses were conducted by Field Training Section, with 217 personnel trained over 31 courses in September 1990. Technical assistance was provided to 82 Wing, 1 and 6 Squadrons. Adventure training was conducted, with activities covering leadership, team exercises, physical training, map reading and navigation being carried out. In April 1990, the Squadron took part in Exercise 'Cope Thunder', followed by the 'RAM90' exercise in the United States in July 1990.

The 482 Squadron team won the Air Vice Marshal Hewitt trophy in May 1991 after competing in a RAAF-wide competition for rifle shooting.

On 2 December 1991, 482 Squadron and 3 Aircraft Depot were amalgamated to form 501 Wing, with final disbandment of the Squadron to be completed by 30 June 1992. Before disbandment took place, 482 Squadron celebrated its golden jubilee on 16 March 1992 with a parade, static display and all ranks ball. Five ex-members of 4 Repair and Salvage Unit (4RSU), from which 482 Squadron had been formed, handed over the 4RSU banner, to be retained as part of the 482 Squadron history.

Commanding Officers

10 May 1946	—Squadron Leader J.E. Jackson
14 December 1948	—Wing Commander W.I. Matson
3 January 1957	—Wing Commander J.E. Jackson
30 March 1961	—Wing Commander C.W. Butcher
3 December 1962	—Wing Commander D.L. Drysdale
6 January 1964	—Wing Commander I.McK. Keir
25 July 1968	—Group Captain J. Hughes
3 January 1973	—Group Captain E.J. Whitehead
18 December 1975	—Group Captain I.T. Sutherland
19 December 1977	—Group Captain M.R. Wade
February 1979	—Group Captain W.M. Collins
12 January 1981	—Group Captain J.N. Spencer
27 May 1983	—Group Captain N.E. Cooper
16 January 1986	—Group Captain P.D.C. Watson
5 July 1988	—Group Captain K.R. Webber
13 November 1991	—Group Captain R.T. Holsken

486 MAINTENANCE SQUADRON

Strength and Achievement

CHRONOLOGY

26 August 1946	—formed at Schofields, NSW
22 June 1949	—moved to RAAF Richmond, NSW
20 April 1954	—moved to Canberra, ACT
29 August 1958	—moved to RAAF Richmond
3 August 1964	—Squadron disbanded
28 March 1966	—reformed at RAAF Richmond
November 1994	—based at RAAF Richmond, NSW

NARRATIVE

486 Maintenance Squadron was formed at Schofields, New South Wales, on 26 August 1946 with an establishment of 10 officers and 247 other ranks under the command of Wing Commander A.J. Abicair.

On 22 June 1949 the unit moved to RAAF Richmond. During the 1950s the Squadron maintained Dakota aircraft, undertaking engine changes, engine overhauls and installation of long range fuel tanks, spraying equipment and electrical power in aircraft. Some Wirraway aircraft were also overhauled. On 22 August 1951, a Canberra bomber arrived and the unit was stood down to attend the taking over of the bomber.

On 20 April 1954 the Squadron moved to Canberra and became responsible for servicing of two Convair Metropolitan aircraft in addition to the Dakotas. On 29 August 1958 the Squadron returned to RAAF Richmond in preparation for the arrival of the new C-130A Hercules.

On 3 August 1964 the Squadron was officially disbanded—personnel, equipment and technical records were equitably distributed between

36 and 38 Squadrons. On 28 March 1966, 486 Squadron was reformed under Squadron Leader C.M. Bevan at RAAF Richmond.

On Christmas Eve 1974 Cyclone *Tracy* struck Darwin and from then until New Years Eve a maximum effort by Squadron personnel pushed 19 C-130s into the air bound for Darwin.

The 40th Anniversary of the Squadron was held on 26 August 1986 and in October 1988 RAAF Base Richmond hosted the Australian Bicentennial Air Show. 486 Squadron was given the task of supervising, co-ordinating and assisting tarmac movements and Squadron personnel found themselves servicing aircraft never seen before in Australia.

In the later years of the Squadron's history its role has been to undertake all C-130 Hercules aircraft servicing up to Depot Level Maintenance, as well as Boeing 707 aircraft servicing including flight simulators and all supporting ground service equipment. The Squadron still operates at RAAF Base Richmond.

COMMANDING OFFICERS

22 August 1946	—Wing Commander A.J. Abicair
1 December 1949	—Squadron Leader F.M. Timms
10 July 1950	—Flight Lieutenant F.T.J. Sellars
14 August 1951	—Squadron Leader J. White
19 March 1956	—Squadron Leader G.S. Smyth
27 October 1958	—Wing Commander G.F. De La Rue
9 January 1961	—Wing Commander F.A. Cousins
1 March 1962	—Wing Commander C.J. Leopold
28 March 1966	—Squadron Leader C.M. Bevan
6 February 1967	—Wing Commander J.C. Kane
August 1967	—Wing Commander E.R. McGeehan
17 February 1969	—Wing Commander E. Harrison
13 April 1970	—Wing Commander C.M. Bevan
15 January 1973	—Wing Commander E.J. Bushell
17 December 1974	—Wing Commander T.J. McGee
5 January 1977	—Wing Commander R.A. Budd
4 January 1979	—Wing Commander E.B. Watson
November 1981	—Wing Commander N.J. Conn
June 1984	—Wing Commander K.A. Pratt
January 1987	—Wing Commander E. Brown
October 1988	—Wing Commander R. Pryke
January 1992	—Wing Commander T.A. Roediger

491 MAINTENANCE SQUADRON

CHRONOLOGY

20 October 1950	—formed at Iwakuni, Japan
13 December 1954	—Squadron disbanded in Japan

NARRATIVE

491 Maintenance Squadron was formed on 20 October 1950 at Iwakuni, Japan. Squadron Leader C.W. Butcher was the first Commanding Officer.

While the Squadron was stationed in Japan, it provided offshore maintenance for aircraft in the Korean conflict. It also provided personnel to be despatched to Korea on a short-time basis to carry out maintenance. The official role of the Squadron states: 'The function of the Squadron is to carry out servicing up to and including major inspections with the exception of daily servicing and repairs arising therefrom for 91 (Composite) Wing'.

In addition to the Japanese location, the Squadron had a section which was attached to 77 Squadron. Originally it was stationed at Pusan, Korea; however as 77 Squadron moved, the servicing section followed and was subsequently located at places such as Pohang and Hamhung in Korea.

During 1950, various personnel were either posted back to Australia or to Japan and thence to Korea. Personnel also took part in short-term maintenance in Korea—for example, on 2 November 1950, Sergeant Malin and Leading Aircraftman Leslie departed for Korea and carried out an engine change on 77 Squadron Mustang aircraft A68-765.

On 21 February 1951, the first shipment of Meteor aircraft arrived on the aircraft carrier HMS *Warrior* from England. A second shipment of Meteors arrived within the following month.

On 1 July 1951, the station's disaster plan went into phase 3 for Typhoon 'Kate'—all aircraft and buildings were made secure with no damage resulting from the night's heavy rain and stormy winds.

In addition to the maintenance of aircraft, the Squadron also carried out salvage operations of numerous crashed aircraft. For example, on 16 June 1952 a Wirraway flown by Sergeant J.D. Codd was carrying out a low flying exercise when it struck power lines and subsequently crashed into the Hiji River near the town of Nagahama. Sergeant Codd sustained a fractured leg, cut arm and fractured fingers. His passenger, Corporal

R.J. Waddell, received serious head injuries. Japanese nationals who witnessed the crash rescued the airmen from the sunken aircraft, and the injured airmen were later evacuated to the 141st United States Army General Hospital. Waddell died shortly afterwards and was buried in the British section of the Yokohama War Cemetery.

On 1 April 1952 the United States Air Force took over control of the station and a handover ceremony was carried out in the afternoon. The station was also closed on 30 May 1952 due to the likelihood of communist disturbances in the Iwakuni area. With the American takeover of the station obvious social differences surfaced as shown by a quote from the Unit History Record of 25 October 1952:

> The American personnel on this base played their first football game of the season today, called Gridiron. The Australian personnel gave much criticism to this game and after seeing it, the vote was that they much preferred the Australian type of football.

By 13 December 1954, the Squadron had been disbanded at Iwakuni. It is worth noting that the Squadron was formed and disbanded in Japan and thus never set foot on Australian soil.

Commanding Officers

20 October 1950	—Squadron Leader C.W. Butcher
9 March 1951	—Squadron Leader C.J. Leopold
19 December 1952	—Squadron Leader A.H. Campey
30 March 1954	—Wing Commander E.K. De Cean

492 MAINTENANCE SQUADRON

Support for the Lion

CHRONOLOGY

1 July 1977	—formed at Edinburgh, SA
November 1994	—based at Edinburgh, SA

NARRATIVE

492 Squadron was formed at RAAF Base Edinburgh, South Australia, on 1 July 1977, as the maintenance squadron for 92 Wing. Squadron Leader R.J. Stevenson was appointed as temporary Commanding Officer and the Squadron was made up of 206 personnel from the individual maintenance flights of 10 and 11 Squadrons. Aircraft establishment included 10 P-3B and 10 new P-3C Orion aircraft, to replace the ageing SP2H Neptunes.

The unit was established with the responsibility for all aspects of Orion maintenance. Courses were run in a variety of fields and included electronics, avionics, weapons systems, radios, armament and airframe systems. Maintenance personnel from the Squadron were often attached away with 10 and 11 Squadrons to maintain the aircraft. Destinations included Butterworth, Diego Garcia, Cocos Islands, Learmonth, Darwin and New Zealand.

In April 1978, A9-293 became the first 492 Squadron aircraft to reach 5000 flying hours on Exercise 'Rimpac 78'. In October of the same year, and again in 1979 and 1993, the Squadron provided support for 11 Squadron, who won the Fincastle Trophy.

In June 1993, 492 Squadron won the Duke of Gloucester's Cup, having been judged the best operational squadron in the RAAF.

Current 492 Squadron tasks are to:

- support the maritime aircraft fleet and its associated avionics, engines and airframe equipment through all levels of maintenance up to and including total repair;
- maintain all ground support equipment used to support P-3C operations, as well as all base motor transport, aircraft handling equipment and the runway arrester system;
- conduct courses for all trades involved in P-3C maintenance;
- support P-3C deployments in Australia and overseas.

Today, 492 Squadron is the largest maintenance squadron in the RAAF, with about 590 personnel from 23 different trades and specialisations. The Squadron is responsible for maintaining 19 P-3C Orions and associated equipment worth over $1.5 billion, with an annual operating budget for spares and equipment of over $12.5 million.

Commanding Officers

1 July 1977	—Wing Commander C.J. Wetherall
20 December 1979	—Wing Commander J.C. Foster
29 March 1982	—Wing Commander A.K. Jaugietis
9 July 1985	—Wing Commander C.McK. Hingston
7 December 1987	—Wing Commander E.P. Holzapfel
18 December 1989	—Wing Commander K.E. Knight
6 December 1991	—Wing Commander K.J. Drover
12 July 1993	—Wing Commander E.C. Hodges

1 MOBILE PARACHUTE MAINTENANCE UNIT

CHRONOLOGY

7 August 1943	—formed at Richmond, NSW
19 March 1944	—moved to River Farm bivouac, Richmond
29 March 1944	—returned to Richmond
10 May 1944	—moved to Mareeba, Qld
30 October 1945	—disbandment completed

NARRATIVE

1 Mobile Parachute Maintenance Unit (1MPMU) was formed at Richmond, New South Wales, on 7 August 1943, with Flight Lieutenant A.B. Watkins posted on attachment to perform administration duties from 13 August 1943 to 8 September 1943. Flight Lieutenant J.H. Chute was then posted in to perform administrative duties until the arrival on 7 January 1944 of Flight Lieutenant J.A. Milne, who was to command the Unit from 3 January 1944.

From 19 to 29 March 1944, the Unit was located at River Farm bivouac, Richmond, before moving to Mareeba, Queensland, in May 1944. The advance party left Richmond by train on 10 May, arriving in Mareeba on 19 May, followed by the main party who arrived at Mareeba on 28 May. The Commanding Officer, Flight Lieutenant Milne, was granted the rank of acting Squadron Leader before leaving Richmond on 21 May, arriving in Mareeba on 24 May.

The Unit equipment, comprising one heavy tender, two jeeps, one weapon carrier, two motorcycles plus general equipment, left Richmond by goods train with a non-commissioned officer and two airmen as escort, on 9 May, arriving in Mareeba on 19 May 1944.

One weapon carrier and a supply of statichutes departed Richmond on 18 May with two airmen as escort, and arrived in Mareeba on 30 May.

On 31 May 1944, the party was cleared by the medical officer, Richmond Station, as being fit for service anywhere in Australia. Inspection by the medical officer, 5 Squadron Mareeba, also revealed a high standard of Unit health.

As no buildings were available at the American camp site, Mareeba, a suitable site for the Unit was selected on 24 Operational Base. Construction commenced on a wooden building to be used as a packing room and two drying rooms, together with headquarters and orderly room. Three stores tents housed the equipment section, parachute maintenance and parachute inspection sections. Transport and barracks occupied a small marquee. Construction also began on a bush timber building for use as a recreation room.

In May 1944, an application was made to Regional Canteen Services for a grant of £100 ($200) to establish a canteen, which opened on 14 June 1944. On 30 June, construction of the Airmen's Mess, kitchen and Sergeants' Mess was completed.

Members of the 1st Australian Parachute Battalion visited the Unit from 20 to 27 June 1944 for harness fitting and adjustment training. On 22 June, three members of the Unit proceeded to Richmond to pick up and act as escort for 'x' type statichutes. The Australian Army Service Corps (AASC) 6th Australian Division AIF visited the Unit on 3 July 1944 for parachute and supply dropping instruction.

The Commanding Officer of 1MPMU consistently reported morale as excellent, but a statement by the 5 Squadron medical officer rates a mention. He notes: 'The Unit as a whole is healthy, but contains a remarkably high number of neurotics, possibly due to the fact that this is the first time any member of the Unit has lived under field conditions'.

At the end of October 1944, a mobile parachute maintenance hut arrived by train from 1 Divisional Works Office Sydney. Assembly of the hut began on 24 October, with an expected completion date of 3 November.

During November 1944, the Unit welfare tent was inaugurated, providing reading and writing facilities. It was also reported that RAAF Headquarters welfare had been requested to purchase a piano for the Unit.

Over the festive season, the Unit was well entertained: personnel saw the *Rats of Tobruk* at the nearby Army Unit, held a card party on 11 December, a dance on 15 December and another on New Years Eve. The Presbyterian padre conducted a church parade on 12 January 1945.

On 3 March 1945 the Unit weapon carrier was involved in an accident which resulted in one man being killed and four injured.

From 20 to 26 March, folding of parachutes took place in preparation for a live drop exercise by 1st Australian Parachute Battalion, also involving members of 1MPMU. Personnel participating in the jump were lectured by Squadron Leader Milne, and undertook ground training on 27 March. Parachute jump 'Exercise Bird' took place from 6 to 9 April 1945. Prior to this, on 3 April, 6 NEA Initial Parachute Folding and Maintenance Course commenced.

On 28 April 1945, the tennis court constructed by Unit members was officially opened. Numerous social activities were organised around this time, including a well-attended and entertaining quiz evening in the Airmen's Mess on 7 May, a musical evening provided by the RAAF String Quartet on 14 May, and a sports meeting to celebrate the first anniversary of the arrival of the Unit. On VE Day, 9 May 1945, a thanksgiving service was held, followed by a Unit stand-down.

The entertainment continued with a crib and euchre evening on 21 May, with the arrival of the piano from NEA welfare section on 24 May. Inoculations and vaccinations given on 3 June by the medical officer of 1st Australian Parachute Battalion did not slow things down, as musical entertainment was provided by two Unit members on 4 June, followed by a quiz session. A basketball game against 1st Australian Parachute Battalion on 10 June resulted in a win for 1MPMU. A Unit dance in the form of a log fire cabaret was held on 28 June, and there were weekly picture shows at 24 Operational Base. The tennis court continued to be popular until July, when there were no more tennis balls available.

Parachute folding commenced on 30 and 31 July in preparation for the live drop exercise 'Parapak 11', involving members of 1st Australian Parachute Battalion. On the first day 160 parachutes were packed, and packing was completed on 3 August. The live drop was held over several days, ending on 7 August.

The Unit stood down on the day of the Japanese surrender on 15 August 1945, and a camp fire cabaret and dance were held on 31 August. A boost was given to morale during September with the attendance of the Unit at a bivouac at Mossman, where the mayor of Mossman officially welcomed two parties of personnel.

On 1 October, the Unit commenced disbandment, with the piano being returned to the welfare section and equipment to various stores units. The final party departed from the Unit by tender for Townsville on 30 October 1945.

Commanding Officer

3 January 1944 —Flight Lieutenant J.A. Milne

3 MOBILE TORPEDO MAINTENANCE UNIT

CHRONOLOGY

1 March 1944	—formed at Cairns, Qld
April 1944	—moved to Darwin, NT
September 1945	—last entry in Unit History Record

NARRATIVE

Formed out of a section of 1 Torpedo Maintenance Unit at Cairns, 3 Mobile Torpedo Maintenance Unit came into existence on 1 March 1944. Commanded by Flying Officer R.A. Jones, the small unit (which consisted of one officer, 23 airmen and one United States Navy torpedo specialist temporarily on loan) was to inspect and maintain torpedoes in storage pending their use by operational squadrons.

In April the Unit embarked in the *R.P. Warner* and, taking its newly received technical equipment and an allocation of torpedoes, sailed for Darwin. Most of the following month was spent setting up maintenance areas and workshops while additional stores, including stocks of practice torpedoes, were received.

On 25 June 1944, 3 Mobile Torpedo Maintenance Unit was called upon to perform its first 'operational' task when it supplied two live torpedoes to 43 Squadron. Catalina A24-45 was to conduct a strike with these weapons.

The Catalinas of 20 and 43 Squadrons, which were supported by the Unit, utilised their flying boats in the night mine-laying role, and made little use of the torpedo as an offensive weapon. To assist in the mine-laying campaign the Unit joined with the United States Navy in conducting trials to see if torpedo lifting slings could be used to load American mines onto the Catalinas. These trials were successful and the slings were subsequently used for this task.

By December, a torpedo overhaul program had been instituted to maintain these complex weapons in a serviceable condition in the harsh tropical environment. December also saw the arrival of a new Commanding Officer, Flying Officer N.W. Allan.

During 1945 the Unit continued its routine duties with one major difference—torpedo training lectures were instituted for the aircrews of both 20 and 43 Squadrons. This was supplemented by the issue of practice torpedoes to both squadrons.

In August 1945, Flying Officer J.H. Evans became the new Commanding Officer. Throughout September the Unit's equipment was allocated for storage and personnel were posted out for discharge.

3 Mobile Torpedo Maintenance Unit disbanded, probably in September 1945.

Commanding Officers

March 1944	—Flying Officer R.A. Jones
December 1944	—Flying Officer N.W. Allan
August 1945	—Flying Officer J.H. Evans

1 REPAIR AND SALVAGE UNIT

CHRONOLOGY

18 March 1942	—formed at Daly Waters, NT
19 June 1942	—moved to Manbulloo
24 June 1943	—arrived at Higgins Field, Cape York, Qld
4 January 1945	—Headquarters opened at Mount Druitt, NSW
15 May 1945	—arrived at Morotai
17 June 1945	—landed at Labuan
26 December 1945	—Unit disbanded

NARRATIVE

1 Repair and Salvage Unit was formed at Daly Waters on 18 March 1942 and on 6 April 1942 Flying Officer Woolley arrived as temporary commander. During May 1942, the stores and transport sections of the Unit were formed in Melbourne and Squadron Leader N.B. Tamlyn was posted in as the Commanding Officer on 16 June 1942.

The advance party of the Unit, consisting of Pilot Officer Harmer and 18 airmen, left Melbourne and arrived at Katherine, Northern Territory, on 19 June to set up the Unit's camp at Manbulloo. The Daly Waters element became a detachment of 1 Repair and Salvage Unit.

The main party left Melbourne on 13 July, arriving at Alice Springs on 23 July. The Unit formed a convoy of 72 vehicles—the largest convoy to move along the north–south road to date—and arrived at Manbulloo three days later.

Makeshift stores and transport sections were established alongside the workshops section, as the Unit was to re-locate to Katherine as soon as the camp there was established. During September 1942, the final elements of 1 Repair and Salvage Unit moved from Daly Waters to Katherine, leaving behind the stores section which had expanded as the North-Western Area Stores Pool; this was handed over to 9 Stores Depot on 23 September 1942.

The onset of the wet season during November resulted in several stores marquees being blown down during local storms. Strong willy-willies swept through the headquarters and stores orderly room tents twice without warning. Also during November, work commenced on repairing and overhauling Wright Cyclone engines. This was severely hampered by the lack of machine and hand tools, as well as having to carry out the work in open tents.

On 1 May 1943, 1 Repair and Salvage Unit was advised of a move. Operations ceased on 3 May and the advance party left Manbulloo in convoy on 30 May 1943 for Darwin. The first train-load of stores and equipment left Katherine for Darwin on the same day, followed by the second on 1 June, the third on 2 June and last one on 3 June.

The MV *Wanaka* arrived in Darwin on 8 June and commenced loading the Unit's equipment the next day. *Wanaka* arrived off Red Island Point (on the western tip of Cape York Peninsula) on 15 June. A temporary camp was set up at a site occupied by 28 Operational Base Unit.

Work commenced on the permanent camp at Higgins Field airstrip. Apart from the main Unit at Higgins Field, two detachments were formed, one at Thursday Island and the other at Merauke. A daily ferry service was inaugurated between Higgins Field and Horn Island for resupply and administration.

An alert was sounded on 30 August 1943 when a suspected Japanese reconnaissance aircraft flew over Higgins Field.

Construction of the camp continued steadily. On 5 March 1944 the cinema commenced operating and construction of the transport buildings began on 12 June 1944.

The Horn Island detachment was disbanded on 21 August 1944 and all personnel returned to Higgins Field.

Tragedy struck the Unit on 22 August 1944 when Norseman aircraft A71-1 force-landed in the sea with four members of the Unit aboard. Leading Aircraftman C.D. Searle was killed and Corporal V.C. Robinson, Leading Aircraftmen D.R. Tomasetti and L.J. Hart spent 15 hours in the sea before being rescued.

A large bushfire swept down on the Unit from the north-east on the afternoon of 6 October 1944. All transport was immediately mobilised to move the petrol dump out of the fire's path and all personnel were called out to fight the fire with beaters.

An outbreak of smallpox occurred in the area in October, causing the whole Merauke, Thursday Island, Horn Island and Higgins Field area to be quarantined.

On 5 December 1944 advice was received that the first 100 members were to leave Higgins Field by air the next day for Townsville. Personnel left Higgins Field in small parties, mostly by air; the 50-man rear party left Red Island Point on 17 December to embark on SS *Islander* which arrived at Sydney on 30 December.

The administration offices opened on 4 January 1945 at Mt Druitt Airfield. On 9 January, the Unit received advice that it was to prepare to move with 86 Wing. In preparation the technical members of the Unit were attached to 2 Aircraft Depot to undergo courses on Mosquito and Beaufighter aircraft.

On 29 January 1945, 26 airmen left Sydney for Meeandah; a convoy followed on 19 February. On 27 March, personnel, stores and equipment left Meeandah. This was followed by the advance party which proceeded from Mt Druitt on 20 April. Others left by rail for Brisbane on 29 April. 'Echelon B' sailed from Brisbane on the *Simon Bamberger* on 2 May. The last members left Mt Druitt for Brisbane on 8 May, from where, as 'Echelon C', they sailed on the *Sea Ray* on 11 May.

The *Simon Bamberger* arrived at Morotai on 15 May, to prepare for the Labuan landing. On 5 June the Unit left Morotai in convoy and arrived at Labuan on 17 June 1945, though it was not until 7 July that the area was cleared and Unit personnel were able to move to their permanent camp site. At the end of the War the Unit strength totalled 330.

1 Repair and Salvage Unit's workload increased over the months following the Armistice, preparing some aircraft and vehicles for return to Australia and others to go to Japan with 81 Wing.

The Unit ceased operations on 8 December and was disbanded on 26 December 1945.

Commanding Officers

16 June 1942	—Squadron Leader N.B. Tamlyn
26 January 1943	—Squadron Leader M.S. Walker
5 November 1943	—Squadron Leader K.R. Dalziel
30 August 1944	—Squadron Leader V.H. Johns

2 REPAIR AND SALVAGE UNIT

CHRONOLOGY

1 December 1944	—formed at Mt Druitt, NSW
1 January 1945	—re-named 2 Repair and Servicing Unit
15 October 1945	—Unit disbanded

NARRATIVE

2 Repair and Salvage Unit was formed at Mount Druitt, Sydney, on 1 December 1944. Its name was changed to 2 Repair and Servicing Unit on 1 January 1945.

The first major task for the Unit was to dismantle a Beaufort aircraft held by 2 Aircraft Park and transport it by road to the Beaufort Division at Mascot.

On 6 January 1945 the Unit assumed responsibility for the general administration of Mt Druitt Station. This included the barracks, messing, guards, duty rosters and such.

During the short life of the Unit personnel dismantled, repaired, salvaged and serviced the following types of aircraft: Spitfire, Beaufort, Dakota, Anson, Wackett, Oxford, Boomerang, Wirraway, DH Dragon, Beaufighter, Stinson and one Seafire. The Royal Navy Seafire repaired was S139. It was being flown by a Pilot Officer Daniels on a cross country flight from the Royal Navy base at Schofields on 3 July 1945 when, shortly after take-off, the controls failed to respond and Daniels decided to bale out. When the canopy was released it struck the pilot on the head, knocking him unconscious. When he came to, he found he was flying over the Mt Druitt airfield and successfully landed. On inspection the fault was discovered to be a damaged aileron.

2 Repair and Servicing Unit repaired the aileron, allowing the Seafire to take off and return to its base (apparently on the same day). Pilot Officer Daniels returned to Schofields in a Norseman aircraft which had been searching for him.

On 8 August 1945 RAAF Headquarters directed that 2 Repair and Servicing Unit was to cease functioning with the view to disband. This was followed by an Operational Order which was received on 16 August. On that same day, the first of 93 postings out was received.

At the time of disbandment, the Unit strength was six officers, five warrant officers, four flight sergeants, 15 sergeants, 34 corporals and 191 airmen, making a total of 255.

The Unit was officially disbanded at Mt Druitt on 15 October 1945. The 23 personnel remaining at that time were attached to 2 Aircraft Depot pending their final posting.

Commanding Officer

1 December 1944 —Flight Lieutenant V.H. Goldfinch

3 REPAIR AND SALVAGE UNIT

CHRONOLOGY

27 May 1944	—formed at Aitkenvale, Townsville, Qld
21 November 1944	—Unit disbanded and used to reform 84 Squadron

NARRATIVE

3 Repair and Salvage Unit was formed at Aitkenvale, Townsville, Queensland, on 27 May 1944. On the previous day, Flight Lieutenant B.R. Cooper had arrived as the temporary Commanding Officer, and by the end of that month the Unit had 258 personnel on strength. The Unit's first permanent Commanding Officer was Squadron Leader F.J. Allen, who arrived to take up command on 3 June 1944.

An interesting aircraft accident occurred on 19 June 1944 which had a happy outcome for all members of the Unit. Kittyhawk A29-704 was landing on the Aitkenvale airstrip to have a belly tank fitted when it struck and killed a steer. After seeking legal advice from North-Eastern Area Headquarters, Unit personnel took the steer to the Unit's messing section, where the cooks butchered and served it up. The Unit History Record notes: 'the fillets of steak proved very tasty morsels!' Incidentally, the Kittyhawk only suffered a damaged port flap and keel cowl fairing. These were repaired, allowing the aircraft to fly again.

During the early days, the Unit was training in preparation for active war service. Training exercises carried out during this time were a 16-kilometre route march by 40 personnel to Harvey's Range Road on 23 June, taking part in active patrols to Mt Stuart (where they defended their positions against attackers, both sides using live ammunition for realism!), 40 personnel taking part in a barge landing exercise on 27 June, and approximately 50 members participating in embarkation/disembarkation exercises using landing barges on 30 June.

The practice landing exercise on 27 June was very nearly a major disaster. While the barge was some distance out to sea, one of the airmen decided to oil the windlass lowering the ramp to improve its efficiency. Unfortunately, he pulled out the wrong pin, and everyone watched in amazement as the ramp disappeared below the water. Fortunately, there were no casualties.

A repair and salvage unit at Noemfoor

A 17 Repair and Salvage Unit convoy with salvaged Vengeance, Giralia

482 Maintenance Squadron members discuss F-111C engine maintenance, March 1981

21 Squadron members work on rebuilding a Vampire trainer, February 1993

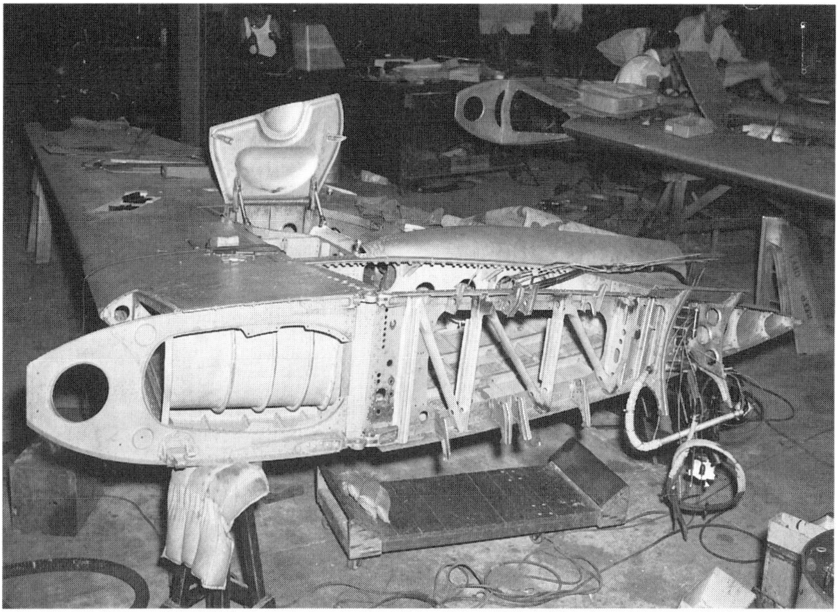

Maintenance is not always routine. A redesigned air intake is fitted to Vampire A78-2, February 1950

9 Squadron personnel work on an Iroquois helicopter, May 1965

First line maintenance on an Orion aircraft at Edinburgh, July 1989

Propeller maintenance on an Orion, July 1989

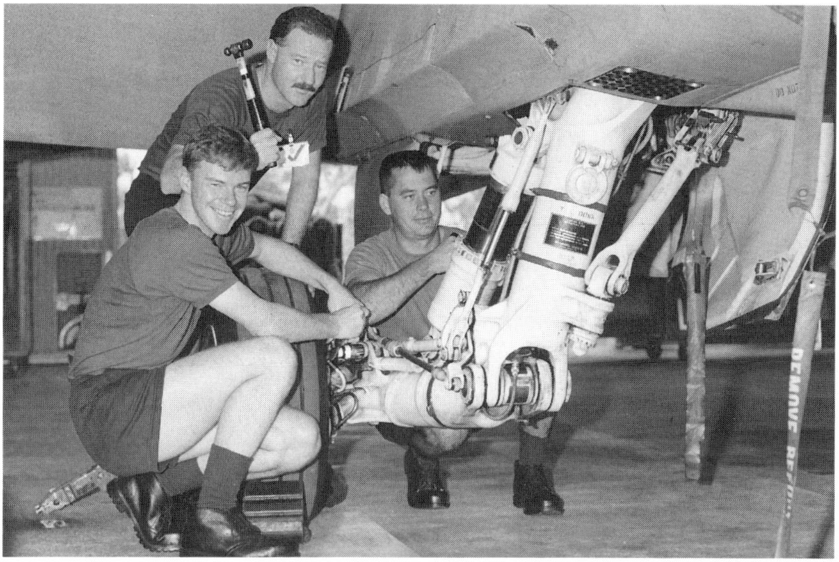

An F/A-18 Hornet undercarriage being serviced, July 1991

An aerial view of 2 Aircraft Park, Geelong, circa 1942

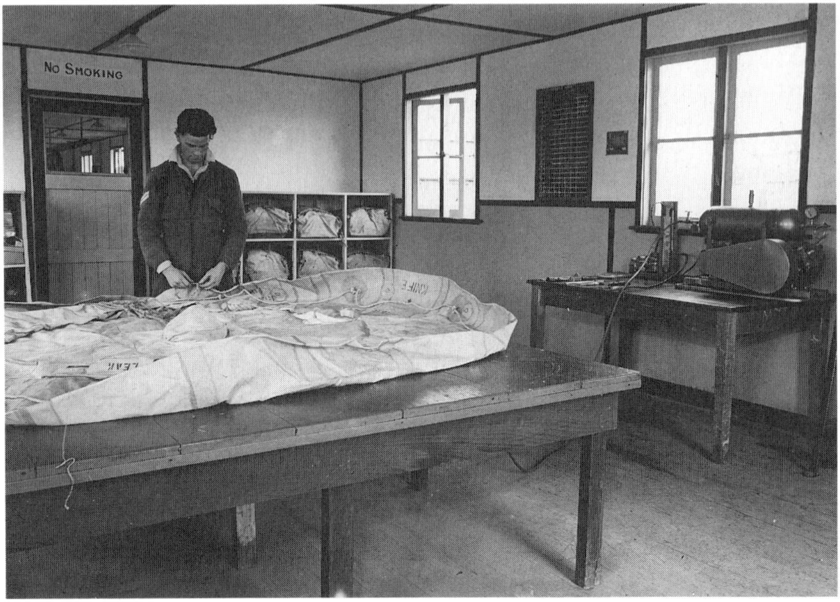

2 Air Observers School Dinghy Maintenance Section, Mount Gambier, May 1945

Tool boards are not new. 2 Air Observers School, May 1945

An F/A-18 undergoes running repairs at 481 Wing, Williamtown

Maintenance personnel of the future. Warrant Officer Dennis Olsen explains the workings of a Merlin engine to Matthew Richards and Claire Stewart at the Queensland Museum

Members of the Unit took part in another major exercise on 11 July 1944, this time a night attack against the Australian Army as the 'enemy'. After boarding landing craft at the wharf, they sailed to Rose Bay accompanied by RAN escort vessels where they stormed ashore at 2250 hours and proceeded to 'wipe out' the 'enemy' manning the defences. 82 Wing's aircraft provided overhead air cover for the assault. After 'destroying' gun emplacements, the raiding party fought their way back to the beach and withdrew to sea in the landing craft. The exercise came to an end with the return to base at about 0200 hours.

Cattle wandering through the area continued to be a nuisance during this time and at last the Commanding Officer decided to do something about it. On 8 July 1944, 18 head of cattle were rounded up, and the local pound was notified to come and collect them. The pound keeper promised to send someone over straight away, and shortly afterwards a stockman arrived to pick up the cattle. At this time, the Commanding Officer decided that to ensure that everything was done by the book a receipt should be obtained from the stockman, who duly signed an acquittance for the cattle and drove them away. However, shortly afterwards another stockman arrived from the pound for the cattle. It seems that the first stockman was actually the owner of the cattle!

Up to this time, the Unit was thought to be at Aitkenvale on a temporary basis. However, on 18 August 1944 an inspection was carried out with the view of determining the camp's suitability for prolonged occupation.

During the latter part of 1944, intensive combat training eased somewhat and the Unit began to take an active part in the various sports activities. This was particularly so for cricket, tennis and swimming. 3 Repair and Salvage Unit also participated with considerable success in the Wing's open sports day on 13 September 1944.

Advice was received on 19 November 1944 that the Unit was to cease operations forthwith with the view of disbandment. Three days later on 21 November the Unit was further advised that personnel and equipment were to be used to reform 84 Squadron. Accordingly, the first 104 postings out were received on 22 November.

Commanding Officer

3 June 1944 —Squadron Leader F.J. Allen

4 REPAIR AND SALVAGE UNIT

CHRONOLOGY

1 June 1942	—formed at Laverton, Vic.
20 October 1942	—advance party to Pell Field
1 January 1945	—re-named 4 Repair and Servicing Unit
May 1945	—moved to Winnellie, NT
15 December 1945	—commenced move to Parkes, NSW
15 April 1946	—moved to Amberley, Qld
10 May 1946	—re-named 482 Maintenance Squadron

NARRATIVE

4 Repair and Salvage Unit was established at RAAF Laverton on 18 May 1942 with acting Flight Lieutenant H.R.P. Relf as the Unit's acting Commanding Officer. After several weeks working on repair and salvage tasks around the Laverton and Werribee areas, the advance party, comprising two officers and 61 airmen moved out on 20 October 1942 to North-Western Area (NWA), installing itself at Pell Field.

For the next three years the Unit remained in the NWA recovering lost aircraft and either repairing them or reducing them to parts. In the course of this activity, many small parties of men, or individuals, spent lengthy periods in the bush away from settled camps, living rough while searching for, and working on, crashed aircraft. The types of aircraft recovered or reduced by the Unit included virtually every type operated by the RAAF in the NWA during this period, and the work required a sound knowledge of the different engineering requirements of the various aircraft involved. Often during these activities, the only link between the temporary camps and the main Unit was the Unit's own aircraft which flew in replacement personnel, needed repair materials, food, water and the accommodation requirements of the salvage team. At Pell Field itself, the Unit constructed an indoor stage and engaged in football matches with other formations in the area, with varying degrees of success. On 5 December 1942 an athletics meeting was held. By 31 January 1943 the Unit comprised 11 officers and 353 other ranks, all of whom became accustomed to 'the wet'.

The Unit engaged in various airfield defence exercises and construction of fire breaks during this period, involving training for preventive measures should the airfield be attacked by the enemy using gas.

On 21 August 1943, the alarm was sounded at 0300 hours, and aircraft were sighted overhead proceeding south at 0315. The aircraft were sighted returning at 0330 and one machine was observed to peel off and come down to attack. It dropped three anti-personnel bombs and an incendiary, causing superficial damage to an empty sentry box. There were no casualties, although telephone communication was temporarily cut.

By the end of November 1943 the Unit comprised 10 officers and 520 other ranks. In January 1944, Squadron Leader Bradley, having completed his tour in the NWA, was replaced by Squadron Leader W.J. Symons. There were many other changes of personnel at this time. In January 1945 the Unit's name was changed to 4 Repair and Servicing Unit and in May the Unit moved to Winnellie and re-established itself there.

From this period on, as war activities moved north away from Australia, the strength of the Unit slowly declined and by the end of August the strength was 11 officers and 385 other ranks.

The Commanding Officer's report for September 1945 commented that the Unit continued to function in a 'smooth running manner' but by October the report lamented the confusion into which reductions and postings had thrown the Unit. The boxing competition was discontinued because of a lack of competitors, but tennis and softball, with their night-lit courts, remained popular activities.

4RSU moved from Darwin on 15 December 1945 by road and rail via Alice Springs, Quorn, Mildura and Tocumwal, to Parkes. Most of the convoy arrived at Parkes between 26 and 29 December, and the Unit moved on to Amberley on 15 April 1946. The last notation in the Unit History Record is a handwritten note indicating that this Unit was renamed 482 Maintenance Squadron on 10 May 1946.

Commanding Officers

1 June 1942	—Flight Lieutenant H.R.P. Relf
November 1942	—Squadron Leader W.A. Bradley
3 January 1944	—Squadron Leader W.J. Symons
28 April 1945	—Squadron Leader J.E. Jackson

5 REPAIR AND SALVAGE UNIT

CHRONOLOGY

27 May 1944	—formed at Bohle River, Townsville, Qld
24 March 1945	—arrived at Morotai
1 November 1945	—Unit disbanded

NARRATIVE

When 86 Squadron was disbanded, the personnel and equipment were utilised to raise 5 Repair and Salvage Unit at Bohle River, Townsville, Queensland, on 27 May 1944. The first task of the Unit was the servicing of 25 Kittyhawk aircraft before delivery to 80 Squadron. The first 12 aircraft were ferried to that squadron on 5 June 1944, and the remaining 13 on 17 June.

In July the Unit took part in an extensive defence training program instituted by Pilot Officer F.P. Cotter, an air defence officer who had defence duties experience in Dutch New Guinea. The course consisted of physical training, rifle, pistol and sub-machine gun shooting at Claden Rifle Range and route marches and bivouacs. During the period 14–17 August, 47 men undertook a route march to Bluewater and return, bivouacking there for four days and undertaking field firing, section exercises, night patrolling and general 'toughening' courses. By the end of September, all members of the Unit had been fully trained.

On 13 and 14 November two Unit vehicles were stolen. The first was a stores tender stolen from Townsville, and the second a weapon carrier stolen from a bivouac site at Alligator River.

Headquarters North-Eastern Area directed that the Unit move to Vidlers Paddock, six miles from Townsville on the main road to Charters Towers, to enable the reformed 86 Squadron to use the airfield. This move did not eventuate but, on 10 February 1945, the Unit was ordered to prepare for movement overseas. Loading of equipment commenced on 27 February, and six officers and 194 men embarked on the *Carlos Carrillo.* The ship departed on 12 March 1945, initially beating through seas so rough that the Unit's Wiles Cooker broke its shackles in the hold and was badly damaged. The troops arrived at Morotai on 24 March 1945.

A camp at Morotai was established in heavy rain and mud up to 2 feet deep. To prevent vehicles from being bogged a corduroy road was built.

The activity and the conditions were the cause of many minor accidents and injuries. However, the weather changed, and the *Herber Green* arrived on 28 March 1945, and unloading commenced.

Six Beaufighter, four Boomerang and two Boston aircraft were allotted to the Unit for repair and, on 5 April 1945, Sergeant McMinn and Leading Aircraftmen King and Grigg travelled to Leyte to repair a Beaufighter. Three days later, the Unit assumed responsibility for the removal of crashed aircraft from the Wama Strip, where the Unit servicing area was located. In addition, Corporal Gardiner and Leading Aircraftmen Cowen and Purdon travelled to Biak to effect temporary repairs to Beaufighter aircraft. The repairs were completed and the aircraft were ferried back to Morotai on 4 May.

A forward detachment was formed at Tawitawi, under the command of Flight Lieutenant W.L. Carson. The detachment members claimed to be the only Australians on the island.

The Unit reached its full strength when the rear party arrived from Townsville on 24 June 1945.

Advice was received on 4 July that 5, 9 and 14 Repair and Salvage Units would combine to form 16 Aircraft Repair Depot. By the 16th, all the officers and men from 5 Repair and Salvage Unit were posted to the depot, with the exception of Flight Lieutenant P.J. Clynes, who remained as the cadre (and only member of) the Unit. This organisation remained until 6 August 1945, when 16 Aircraft Repair Depot was disbanded, and the repair and salvage units reformed.

Disbandment instructions were received on 28 September 1945, and the Unit was disbanded on 1 November 1945.

Commanding Officer

2 June 1944 —Squadron Leader W.J. Waite

6 REPAIR AND SALVAGE UNIT

CHRONOLOGY

10 April 1944	—formed at Mt Druitt, NSW
23 February 1944	—moved to Fenton, NT
30 June 1945	—arrived at Morotai
24 July 1945	—landed at Balikpapan, Borneo
28 November 1945	—Unit disbanded

NARRATIVE

6 Repair and Salvage Unit was raised at Mt Druitt on 10 April 1944. Initially a lodger unit of 18 Repair and Salvage Unit, 6 Repair and Salvage Unit had, by the end of August 1944, taken over the facilities of the former unit.

The first salvage operation recorded by the Unit was the recovery of a Beaufort from Mascot and its transportation to 5 Aircraft Depot Wagga Wagga. Other operations involved the collection of three Wackett Trainer fuselages and six mainplanes from the Newcastle Aero Club, and 240-hour servicing of Boomerang aircraft.

Members of the Unit socialised with the local population, and assisted the community in time of emergency. On 7 November 1944, 30 airmen assisted in controlling a bushfire which was burning 6 miles north of the station. This was repeated on 10 and 11 December, when 50 men fought a conflagration near Penrith and 25 attacked a bushfire near Woodstock.

Plans to move the Unit to a new location were mooted on 17 November 1944. On 13 December, Flight Lieutenant R.H. Barnes and a party of 45 departed by train, bound for Fenton, Northern Territory. The group arrived on 27 December. Unit stores were loaded aboard SS *Helga Moller* at Circular Quay and, accompanied by Flying Officer B.V. McCauley and 11 airmen, arrived at Darwin on 15 February 1945. The main body of 300 officers and men left Mt Druitt by train on 9 February, and the rear party of 40, under the command of Flight Lieutenant P.J. Willington, departed on the 14th. These parties arrived at Fenton on 21 and 23 February 1945 respectively.

The Unit was responsible for the maintenance of Liberator aircraft, and the salvage of other types. Whilst attempting to salvage a Mitchell from an area 12 miles east of Adelaide River on 30 March the salvage party was forced to abandon a weapon carrier which had become bogged, and returned to camp on horseback. The carrier was recovered on 2 April.

Members of the Unit travelled from the 51 Mile camp, at which they had staged from Fenton, to Darwin, on 17 June 1945, where they embarked on the *Luis Arguello*, en route for Morotai. The Unit arrived at Morotai on 30 June 1945 and Flight Lieutenants Barnes and Willington led 49 airmen ashore to prepare a camp for the main body, which landed next day.

One of the features of the Unit's stay at Morotai was participation in an 82 Wing defence exercise, training for the defence of the Balikpapan perimeter. On 12 July 1945, 406 members of 6 Repair and Salvage Unit boarded the *Cleveland Forbes*, and arrived at Balikpapan on the 16th. However Unit personnel did not land until 24 July, after witnessing an air raid by 20 'Betty' bombers during the previous day. The enemy bombers had been engaged by the anti-aircraft defences, and one bomber had been shot down by fighters scrambled from the Sepinang airfield.

A camp site was established between the Vasey Highway and the sea, in dense jungle. Preparation required the bulldozing of an access road and was hampered by cloying mud and air raids. During the third raid, one of the ships suffered a near miss, but the action did not prevent the completion of the camp site on 4 August 1945.

An outbreak of dysentery hampered activity, but did not prevent the salvage of Boomerang and Kittyhawk aircraft from Sepinang. However, on 1 October, 82 Wing advised that the Unit would disband in the near future. The Unit ceased to function on 20 November and was disbanded on 28 November 1945.

Commanding Officers

4 July 1944	—Squadron Leader V.H. Johns
1 November 1944	—Squadron Leader P.J. Allen
27 May 1945	—Squadron Leader A.J. Brown
19 August 1945	—Squadron Leader J. Hearnden

7 REPAIR AND SALVAGE UNIT

CHRONOLOGY

3 June 1942	—formed at Richmond, NSW
13 January 1943	—moved to Pell, NT
10 February 1944	—moved to Darwin Civil Aerodrome
August 1945	—last entry in Unit History Record

NARRATIVE

7 Repair and Salvage Unit commenced to function at Richmond, New South Wales, on 3 June 1942, under the command of Flying Officer J.A. Campbell.

Members of the Unit travelled to areas including Coffs Harbour, Singleton, Grafton, Nowra and Nabiac to salvage Wackett, Battle, Beaufort, Spitfire and Hudson aircraft. The first recorded operation was the conversion of a Douglas DC-2, A30-7.

On 5 September 1942, instructions were received that the Unit would be moving to Port Moresby. This planned move was deferred until the Unit was suitably equipped, and finally cancelled on 22 September.

Pilot Officer A.J. Cole and a party of 23 airmen travelled to 1 Aircraft Depot, Laverton, Victoria, to gain experience on Spitfire aircraft. The party returned to Richmond on 10 November and, on 12 December 1942, notification was received that the Unit was to move to Pell, Northern Territory. On 31 December 1942, the main party embarked on the SS *James Russell Lowell* and arrived at Darwin on 12 January 1943. The Unit disembarked next day, and travelled by road to the camp site at Pell, at which a mobile works unit had erected 43 tents, four mess rooms, two kitchens and four showers.

The Unit was greeted by two air raid alerts, on 20 and 21 January 1943, before commencing duty on the 23rd when 1 Mobile Repair Unit travelled to Daly Waters to repair a Spitfire. The Unit established three mobile repair units and salvage units, which travelled to Pine Creek, Humpty Doo, Cape Shield, Millingimbi, Bathurst Island, Melville Island, Wyndham and Daly Waters. The Unit also maintained and salvaged aircraft from the fighter airfields at Sattler, Strauss and Livingstone. 7 Repair and Salvage Unit maintained Spitfire aircraft, but also serviced Mitchell and Wirraway types. As an example of the quantity of work undertaken by the Unit, during July 1945 the Unit held 26 Spitfire, three Wirraway and two Mitchell aircraft for inspection and repair.

On 5 April 1943, a 'Zeke' crash-landed on Melville Island. Pilot Officer O'Conner and a party departed from Pell to salvage the aircraft, and arrived back, disappointed, on 27 April, as the aircraft had already been recovered.

An advance party departed from Pell, bound for the Darwin Civil Aerodrome, on 10 February 1944. The move was completed on 15 February.

Warrant Officer Cox was killed when Spitfire A58-310 crashed into Darwin Harbour, whilst on a test flight, on 5 October 1944.

Sport was important in maintaining morale. Members of the Unit participated in Australian football, soccer, tennis, athletics, cricket and boxing. Boxing proved very popular and volunteers from the Unit built a stadium next to the Parap Hotel, which was capable of housing up to 5000 spectators at well-organised boxing matches.

The final entry in the Unit History Record was made in August 1945.

Commanding Officers

3 June 1942	—Flying Officer J.A. Campbell
1 February 1944	—Squadron Leader W.I. Matson
March 1945	—Squadron Leader A.L. Goden

9 REPAIR AND SALVAGE UNIT

CHRONOLOGY

15 September 1943	—formed at Werribee, Vic.
29 May 1944	—operational at Sattler, NT
1 February 1945	—arrived at Morotai
24 September 1945	—Unit disbanded

NARRATIVE

9 Repair and Salvage Unit was raised at Werribee, Victoria, on 15 September 1943, and Flight Lieutenant R.A. Savage assumed command of the new Unit next day. At the end of September, the strength of the Unit was 49 all ranks, but by the end of May this had risen to 385.

Salvage operations were undertaken at areas such as Seymour, Flinders Island, Landsborough, Stawell and Gipsy Point. In the latter case the co-operation of Navy authorities at Flinders Naval Depot was required, as diving operations were required to discover the wreckage.

On 8 April 1944, the advance party arrived at Sattler, Northern Territory. Camp was set up at Number 6 site, and the Unit became operational on 29 May 1944. The Unit recovered Spitfire and Wirraway aircraft, but all operations were not successful. On 15 June, Flying Officer H.W. Wrath and two airmen attempted an overland journey to gain access to Spitfire A58-117, which had crashed near Point Blaze, but were thwarted by impassable swamp.

Not all operations were related to salvage operations. On 8 September, the Unit installed an electric flare path at Snake Bay.

A conference at 80 Wing Headquarters on 13 January 1945 preceded the arrival of 9 Repair and Salvage Unit at Morotai on 1 February 1945. The main rear party, which consisted of four officers and 129 other ranks, embarked on the SS *Nicholas J. Sincott* on 25 February and arrived at Morotai on 14 March. This group disembarked on DUKWs for the shore next day, and it was not until 5 April 1945 that the final party arrived from the Northern Territory.

Parties from 9 Repair and Salvage Unit participated in the landings at Labuan ('Oboe 6'), Tarakan ('Oboe 1') and Balikpapan ('Oboe 2'). Flight Lieutenant C.G. Jones was appointed the Commanding Officer of the former detachment, which was to work with 457 Squadron. The

35 personnel departed for Labuan on 1 June 1945, and became operational on 20 June, handling two crashed aircraft that day. Warrant Officer Mason and 33 men operated with 452 Squadron at Tarakan, and Flying Officer L.J. Scaysbrook left Morotai with 33 men for Balikpapan on 4 July. The Tarakan and Balikpapan detachments returned to Morotai in September.

With the formation of 16 Aircraft Repair Depot on 16 July 1945, the adjutant, orderly room flight sergeant and equipment flight sergeant were detailed to disband 9 Repair and Salvage Unit. During August, 85 personnel were posted south at the end of their tropical tour, and the anti-corrosion preparation of equipment was being undertaken before the Unit officially disbanded on 24 September 1945.

COMMANDING OFFICERS

16 September 1943	—Flight Lieutenant R.A. Savage
21 August 1944	—Flight Lieutenant R.W. Ayre

10 REPAIR AND SALVAGE UNIT

CHRONOLOGY

1 September 1942	—moved from Amberley to Lowood, Qld
21 November 1942	—moved to Breddan, Qld
18 January 1943	—moved to Macrossan, Qld
13 February 1943	—arrived at Milne Bay, New Guinea
14 November 1944	—Headquarters arrived at Nadzab, New Guinea
1 August 1945	—operational at Lae
24 November 1945	—Unit disbanded

NARRATIVE

Records for 10 Repair and Salvage Unit commence with the move of the Unit from Amberley to Lowood, Queensland, on 1 September 1942. The Unit was then advised that it would be moving further north, to take over work of a detachment of 12 Repair and Salvage Unit at Macrossan.

Two mobile sections, comprising one officer and 53 other ranks, departed for Breddan, Queensland, on 29 October 1942. This party was established by 5 November 1943, and the main party commenced its move to Breddan on 21 November.

On 5 December 1942, the Unit was warned that it would be required to move by sea to Milne Bay. In preparation for the move, 10 Repair and Salvage Unit moved to Macrossan on 18 January 1943, and the advance party, consisting of the Commanding Officer and nine airmen, departed by air for Milne Bay on 4 February 1943. The main body arrived on 13 February 1943.

The Unit serviced and salvaged Kittyhawk, Beaufort, Oxford and Dakota aircraft. In addition, it undertook monthly inspections of motor transport and, on 1 May 1944, as an additional task, 200 personnel assisted with the salvaging of technical equipment from a flood-affected 17 Stores Depot.

Detachments of 10 Repair and Salvage Unit were established at Kiriwina and Goodenough. The former was raised on 1 March 1944, under the command of Flying Officer A. Saunders, and comprised 54 airmen. The Goodenough detachment commenced operations on 9 July 1944.

There was time for relaxation and sport. During June, the Unit's Australian football team won the local RAAF unit competition.

Instructions were received on 13 July 1944 that the Unit was to move to Nadzab, and form the basis for a new aircraft repair depot. On 24 July, 121 airmen were posted from 15 Aircraft Repair Depot to the Nadzab detachment of 10 Repair and Salvage Unit. On 26 July, the Unit was advised to move as much equipment and personnel as possible from Milne Bay to Nadzab by 14 August 1944. However, even though five Dakota aircraft and the ship *Betsy Joan* were allotted to the Unit for the task, this timing was not met. Headquarters 10 Repair and Salvage Unit closed on 13 November 1944, and re-opened at Nadzab next day. The Unit and its equipment were loaded aboard the USS *Simon Benson* on 14 November and disembarked at Lae on 24 November. They proceeded to Nadzab in December, then moved back to Lae on 1 August 1945.

New messes, kitchens, recreation rooms, showers and other facilities had to be constructed at Nadzab, and this process was repeated when the Unit moved to Lae on 1 August 1945, following a party of 60, led by Flight Lieutenant E.J. Littlejohn, who travelled to the area on 14 April to prepare the new camp site.

The Commanding Officer travelled to Madang on 13 December 1944 for discussion regarding the ultimate establishment of a detachment there.

The Goodenough detachment was withdrawn, arriving back at Milne Bay on 1 March 1945, to wait on movement to Nadzab.

On 18 July 1945, Flight Lieutenant Weir-Smith led 24 airmen to Morotai to form the Forward Recovery Section of 10 Repair and Salvage Unit, which became operational on 21 August 1945.

The Madang detachment was disbanded and absorbed by the parent unit on 2 November 1945. 10 Repair and Salvage Unit was disbanded on 24 November 1945.

Commanding Officers

September 1942	—Flight Lieutenant B.L. de Melnotte
10 June 1943	—Squadron Leader H.W. Porteous
2 November 1944	—Squadron Leader P.G. Clark
14 March 1945	—Squadron Leader J.W. Nankivell

11 REPAIR AND SALVAGE UNIT

CHRONOLOGY

1 December 1942	—formed at Mt Druitt, NSW
19 November 1943	—advance party departed for Coominya, Qld
22 January 1944	—arrived at Nadzab, New Guinea
11 September 1944	—advance party departed for Noemfoor
6 February 1945	—main body departed for Morotai
23 April 1945	—advance party departed for Tarakan
October 1945	—last entry in Unit History Record

NARRATIVE

11 Repair and Salvage Unit was established at Mt Druitt on 1 December 1942. Flight Lieutenant A.C. Kerr assumed command of the Unit on 9 December and the Unit initially occupied 'The Castle' at Centenary House. During March 1943, the members occupied a new camp site.

Salvage parties travelled to Coffs Harbour, Braidwood, Camden, Bulli, Strathpine, Gulgong, Forster, Albion Park, Williamtown and Rockhampton. At one stage, the area of the Unit's salvage operations ranged from Bundaberg to Moruya, servicing and salvaging aircraft types such as Vengeance, Beaufort and a Northrop Delta.

One of the more interesting operations was undertaken by the Goolman detachment, commencing on 31 May 1943, when it departed to recover an Anson aircraft which had crashed on Heron Island.

Not all salvage expeditions were trouble free. On the night of 1 July 1943, Flying Officer Allen was leading a party of 12 airmen down Bulli Pass to the rifle range, when the driver lost control of the tender in which they were travelling, and it capsized. No injuries were incurred.

On 22 October 1943, members of the Goolman detachment—Flight Lieutenant Hartnell and 19 other ranks—departed from Amberley for Nadzab, New Guinea. The balance of the detachment returned to Mt Druitt, before the Unit moved to Bradfield Park and movement further north. On 19 November 1943 Flight Lieutenant W.J. Wynne and 69 other ranks departed for Coominya, Queensland. On the following day, Flying

Officer T. Hallick and a party of 52 proceeded from Bradfield Park to Sandgate, Queensland, to operate as an off-loading party at the Meeandah Siding. Squadron Leader C.R. Taylor led the main body to Coominya on 26 November, where they commenced training in the use of firearms, hand grenades and explosives. In addition, daily route marches were undertaken.

A sergeant and 20 airmen departed from Archerfield to join a works unit at Nadzab, and the Unit moved to the Yeerongpilly Army Camp before embarking for overseas on 17 January 1944. They arrived at Nadzab on 22 January 1944. By the end of February, the Unit was fully operational, servicing Vengeance and Kittyhawk aircraft.

On 28 February 1944, the Unit was warned that it was to move 80 per cent of its strength at three days notice. The move was deferred indefinitely, and a ferry flight was established in May, under Flight Lieutenant Wigglesworth, to fly aircraft to forward units. However, on 9 August 1944, 52 airmen departed to assist 22 Repair and Salvage Unit at Noemfoor.

The Unit moved from Nadzab to Lae on 18 August, preparatory for its move to Noemfoor. However, it was not until 11 September that the advance party of two officers and 50 other ranks departed by sea for that destination, with the main body following during the period 18–19 September.

During November, facilities at Noemfoor were erected, but the Unit was not destined to remain on the island for long. On 13 December, Echelon 'A' departed by sea for Morotai, to be followed by the Salvage and Repair Echelon on 22 December. On 2–3 January 1945, Echelon 'B' was flown to Morotai in 26 Dakota flights, and the main body departed from Noemfoor on 6 February 1945.

The advance party of the Unit participated in the landing at Tarakan, departing from Morotai on 23 April 1945. The remainder of the Unit travelled to Tarakan on Landing Ship Tanks and HMAS *Westralia* during May.

11 Repair and Salvage Unit was active in the area, and one of the last entries in the Unit History Record relates to a trip made by a party from the Unit up the Segayap River during October. After a 130-mile each way journey, Special Reconnaissance Department members were evacuated, and stirring tales of the Dyaks recounted.

The final entry in the Unit History Record was made in October 1945.

Commanding Officers

9 December 1942	—Flight Lieutenant A.C. Kerr
15 April 1943	—Squadron Leader C.R. Taylor
1 February 1945	—Squadron Leader I.M. Keir

12 REPAIR AND SALVAGE UNIT

CHRONOLOGY

18 May 1942	—formed at Charters Towers, Qld
2 October 1942	—moved to Breddan, Qld
5 March 1943	—advance party moved to Macrossan, Qld
16 December 1943	—arrived at Kiriwina
13 May 1944	—disembarked at Tadji, Papua New Guinea
1 January 1945	—re-named 12 Repair and Servicing Unit
29 November 1945	—Unit disbanded

NARRATIVE

12 Repair and Salvage Unit was established at Charters Towers, Queensland, on 18 May 1942, with an establishment of eight officers and 381 airmen. The advance party was accommodated in the town, prior to the camp site, 3 miles out of the town, being occupied on 7 June 1944. Tents and buildings were erected at this site.

The Unit operated a detachment at Garbutt from August, and during September aircraft such as a Lockheed Vega, Hudson, Fox Moth, Kittyhawk, Wirraway, Beaufighter and Oxford were assigned to the Unit for repair. On 14 September, 1 Salvage Party recovered aircraft from the north coast of Queensland and, on 20 September 1943, 2 Salvage Party of 42 personnel moved by sea to Horn Island to maintain aircraft there. An example of the arduous work undertaken by these parties was the salvage of Beaufort A9-157, which had force-landed on the east coast of the Gulf of Carpentaria. On 30 March 1943, the Unit reported that 150 mission personnel and five horses were used to tow the aircraft eight miles to a mission station, where the aircraft was dismantled and the parts transported back to Horn Island in the two mission luggers.

On 2 October 1942, the Unit moved to Breddan, where it was joined by 1 Salvage Party. The new camp site was 15 miles from Charters Towers, and the Unit remained there until 5 March 1943, when an advance party proceeded to Macrossan to prepare a camp for the Unit. The move did not impede activities. Seven personnel travelled to Musgrave on 20 May to salvage damaged aircraft, and a special job was undertaken at Townsville on 10 September 1943, involving Catalina A24-52.

Advice was received on 3 October 1943 that the Unit would be serving overseas. On 9 October, an advance party of 12 departed for Kiriwina. The main body boarded the *Lloyd B. Olson* at Townsville on 10 December 1943, and arrived at Kiriwina on the 16th. The Unit remained at Kiriwina until 29 April 1944. After leaving a rear guard of 87 personnel at Kiriwina, the main body travelled to Oro Bay, then to Finschhafen, and finally to Tadji, landing on 13 May 1944. A temporary camp was erected while 50 personnel remained aboard ship to assist with unloading. An air raid warning greeted the men during their first night ashore.

The military situation at Tadji was not comforting. Defensive pill boxes were under construction inside the camp and, on 20 July, it was decided that substantial additions would be made to these strong points. However, as the military situation improved in mid-August, the Unit was instructed to remain at Tadji and commence operations. It had been mooted that the Unit would join the rear guard, which had been moved to Noemfoor on 25 July—this became a detachment of the Unit, and another was based at Los Negros in August.

A notable salvage operation occurred in September. A party was sent to Kerowaagi, in the Chimbu Valley, to dismantle a crashed Beaufort. The aircraft was in wild country (headhunters still lived there) and had to be packed for shipping. This party was finally withdrawn on 12 December, due to the lack of transport to remove the parts salvaged.

The Unit settled in to a routine, and obtained a bulldozer during November to push crashed aircraft off the strip and thus clear it for aircraft landing low in fuel. The Unit also had reason to complain at the standard of food and welfare items. On 14 October 1944 it was reported that the water supply contained slime and vegetable matter, but it did not prevent the surf beach being used to its full potential and the reel and belt team made several rescues.

During 12–13 December huge downpours flooded the equipment section and a 5kVA power source was burnt out, leading to the rewiring of the whole camp site. Tents developed leaks, and two of the jeeps, bogged in the nearby river, were washed out into the surf. The rain came down again on 28 January 1945, with the tents letting in so much water that some occupants had to sleep in the recreation rooms. February brought another natural extreme, and the aircraft being serviced were covered with portable hangars to reduce the cockpit temperature from 152°F to a mere 98 or 100.

The name of the Unit was changed on 1 January 1945 to 12 Repair and Servicing Unit.

From January to April, the routine of the Unit was unbroken. Surfing was still a popular leisure activity, and the first Australian entertainment party to perform at Tadji did so on 26 February and, in general, domestic conditions improved.

Four airmen disassembled a Japanese aircraft at But on 20 April, but the shortage of personnel at Tadji was affecting the amount of work that could be undertaken by the Unit. On 29 June 1945 there were over 100 personnel at Madang who had been posted to Tadji waiting transport, so their collection became an urgent issue.

With the Japanese surrender on 15 August, and victory church services on the following day, the Unit relaxed. 12 Repair and Servicing Unit wound down, and was finally disbanded on 29 November 1945.

COMMANDING OFFICERS

18 May 1942	—Squadron Leader T.A. Cummins
2 November 1943	—Squadron Leader N.B. Tamblyn
29 March 1945	—Flight Lieutenant R.A. Taylor
22 June 1945	—Squadron Leader G.H. Tourn

14 REPAIR AND SALVAGE UNIT

CHRONOLOGY

21 July 1943	—formed at Mt Druitt, NSW
19 March 1944	—moved to Pell, NT
13 April 1944	—moved to Livingstone, NT
23 May 1944	—moved to Ross River, Townsville, Qld
8 October 1944	—moved to Woodstock, Qld
12 March 1945	—moved to Morotai
30 October 1945	—last entry in Unit History Record

NARRATIVE

14 Repair and Salvage Unit was formed at Mt Druitt, New South Wales, on 21 July 1943. The Unit repaired Wirraway, Beaufort, Tiger Moth, Kittyhawk and Ventura aircraft.

On 19 March 1944, the Unit moved from Mt Druitt to Pell, via Alice Springs. On 13 April 1944, the Unit moved by road transport to Livingstone, Northern Territory. The next move was to Ross River, Queensland, where Unit personnel started an aerodrome defence course. The course included unarmed combat, small arms, route marches and river crossings. From Ross River, the Unit moved to Woodstock and on 12 March 1945 departed from Townsville for Morotai. While at Morotai, the morale of the men improved, due to an active participation in the War.

In October 1945, the Unit was advised regarding disbandment, and the final entry in the Unit History Record was made on 30 October 1945.

COMMANDING OFFICERS

3 August 1943	—Squadron Leader R.H. Foord
16 August 1943	—Squadron Leader N.B. Tamlyn
19 October 1943	—Squadron Leader O. Doutch
August 1945	—Squadron Leader N.S. Lake

17 REPAIR AND SALVAGE UNIT

CHRONOLOGY

19 May 1942	—formed at Pearce, WA
29 June 1942	—moved to Cunderdin, WA
13 July 1944	—moved to Pearce
12 February 1946	—disbandment commenced

NARRATIVE

17 Repair and Salvage Unit was formed at Pearce on 19 May 1942 under the command of Flying Officer R.G. Watt. On 29 June 1942, the Unit moved to Cunderdin, where it took over Halbert's garage for vehicle repairs and Baxter's homestead, 'Burnside'.

Once the Unit became operational various aircraft including Beaufort, Buffalo, Wirraway, Kittyhawk, Anson, Hudson and Brewster were allocated for either repairs or salvage. In 1943, a squadron of Boomerangs was posted to the area, and the aircraft came to the Unit for repair.

In February 1943 two hangars were completed and two canvas hangars were erected at Halbert's garage, for the servicing of vehicles in wet weather.

On 21 May 1943, the Unit had its first air raid alarm when enemy aircraft flew overhead.

The Unit formed a strong bond with the local residents at Cunderdin who were disappointed when the Unit returned to Pearce on 13 July 1944.

While at Pearce the Unit continued with its tasks and commenced to disband on 12 February 1946.

COMMANDING OFFICERS

19 May 1942	—Flying Officer R.G. Watt
21 November 1942	—Squadron Leader A.W. Woolley
7 April 1944	—Squadron Leader H.H. Powell
28 June 1944	—Squadron Leader J.E. Jackson
5 March 1945	—Squadron Leader G.H.P. Bond
September 1945	—Flight Lieutenant F.L. Hammond

18 REPAIR AND SALVAGE UNIT

CHRONOLOGY

20 January 1944	—formed at Mt Druitt, NSW
16 August 1944	—moved to Pell, NT
1 March 1945	—moved to New Britain
10 June 1945	—moved to Morotai
8 July 1945	—moved to Balikpapan
30 November 1945	—last entry in Unit History Record

NARRATIVE

18 Repair and Salvage Unit was formed at Mt Druitt, New South Wales, on 20 January 1944. Squadron Leader J.E. Haigh was posted from 14 Repair and Salvage Unit to command the Unit, which received various equipment and had its personnel instructed in small arms, drill and bayonet fighting in preparation for a move north.

On 26 July 1944, the Unit loaded equipment and vehicles on to the SS *Suva* at Glebe Island, bound for Darwin. Unit personnel departed on 16 August 1944 by train, via Albury, Melbourne and Adelaide to Alice Springs and then by road transport to Pell, Northern Territory.

The headquarters orderly room at Pell started to function in August 1944. The Unit repaired, inspected and maintained Mitchell and Beaufort aircraft. Unit personnel attended ground defence training in readiness for the move to the South-West Pacific.

On 1 March 1945 the Unit moved to New Britain and commenced camp construction. On 10 June 1945 the Unit sailed to Morotai and on arriving disembarked to a temporary camp.

On 8 July 1945, the Unit moved to Balikpapan and continued after the War maintaining aircraft while personnel were posted out of the Unit.

COMMANDING OFFICERS

20 January 1944	—Squadron Leader J.E. Haigh
19 September 1945	—Flight Lieutenant W.A. Sleebs

22 REPAIR AND SALVAGE UNIT

CHRONOLOGY

26 April 1943	—formed at Werribee, Vic.
11 December 1943	—moved to Bohle River, Townsville, Qld
February 1944	—moved to Nadzab
20 March 1944	—moved to Cape Gloucester
25 June 1944	—moved to Noemfoor
1 January 1945	—re-named 22 Repair and Servicing Unit
9 April 1945	—moved to Morotai
20 June 1945	—moved to Labuan
30 October 1945	—last entry in Unit History Record

NARRATIVE

22 Repair and Salvage Unit was formed at Werribee, Victoria, on 26 April 1943, under the command of Flight Lieutenant H.E. Coles.

The Unit's first salvage operation occurred on 14 May 1943. Flight Lieutenant Coles and eight other ranks proceeded to Lakes Entrance, Victoria, to salvage Anson AX425. The Unit salvaged many aircraft, including Anson, Hudson and Beaufort aircraft, from Victoria and surrounding areas.

On 11 December 1943, the Unit moved to Bohle River, Townsville, before deploying to Nadzab, Cape Gloucester, Noemfoor, Morotai and finally Labuan. While on these islands, personnel developed camp sites, and constructed aircraft workshops, buildings and runways, in addition to the Unit's normal operations. While on Noemfoor, personnel working on the airstrip suffered from conjunctivitis, due to the glare and dust from the coral airstrip.

On 11 October 1944, the Unit changed its command from Headquarters 78 Wing to 10 Operational Group Headquarters and on 1 January 1945 the Unit changed its name to 22 Repair and Servicing Unit. With the end of the War, volunteers for an occupation force in Japan were called for and, on 30 October 1945, Japanese language lectures commenced under the direction of Squadron Leader Widrowski, assisted by Major Hayasaki, a Japanese prisoner of war. This was the last entry in the Unit History Record.

Commanding Officers

26 April 1943	—Flight Lieutenant H.E. Coles
25 April 1944	—Squadron Leader G.F. De La Rue
1 August 1945	—Squadron Leader S. Medley

26 REPAIR AND SALVAGE UNIT

CHRONOLOGY

20 August 1942	—formed at Laverton, Vic.
3 December 1942	—moved to Werribee, Vic.
June 1943	—moved to Goodenough Island
7 September 1944	—moved to Noemfoor
December 1944	—re-named 26 Repair and Servicing Unit
2 April 1945	—moved to Morotai
May 1945	—moved to Tarakan
18 November 1945	—Unit disbanded

NARRATIVE

26 Repair and Salvage Unit was formed at Laverton, Victoria, on 20 August 1942, to repair and salvage aircraft, motor transport and other equipment, and to repair, maintain and modify equipment beyond the capacity of operational squadrons. Early activities were related to the preparation of demands for equipment and its collection from 1 Stores Depot.

The Unit began its first salvage job on 14 September 1942, when it dismantled a neon sign on the Garrick Theatre, Melbourne, and recovered approximately 14 tons of structural steel. Salvage of aircraft began in October 1942. Flight Lieutenant Ley and Sergeant Lee travelled to Bairnsdale and Lake Victoria to inspect the Unit's first assignments and, as a result, Sergeant Lee left Laverton with a party of 14 airmen to salvage Beaufort A9-61 at Lake Victoria; they completed the job by 22 October 1942.

On 8 October 1942 the Unit came under the control of 4 Maintenance Group.

On 3 December 1942 the Unit moved to Werribee and then, in June 1943, moved to Goodenough Island, New Guinea, to set up camp. While in this area personnel had small arms instruction and had to perform defence works. The Unit salvaged Beaufort, Kittyhawk, Beaufighter, Spitfire and Boston aircraft. On 29 June 1944, the Unit came under the control of 10 Operational Group. On 7 September 1944 the Unit moved to Noemfoor, via Finschhafen and Biak, on the SS *Sidney H. Short*. During December, the Unit changed its name to 26 Repair and Servicing Unit and, on 2 April 1945, embarked aboard the SS *Clarence Roberts* for Morotai. The Unit reached Morotai on 10 April 1945, before deploying to Tarakan in May. Work suffered due to confrontation with enemy ground troops—at one stage the Unit's advance party was only 100 yards from the Japanese lines.

With the end of the War in August there was little work for the Unit and, on 18 November 1945, it disbanded.

Commanding Officers

20 August 1942	—Squadron Leader S.F. Ley
4 October 1944	—Squadron Leader K.R. Dalziel
30 April 1945	—Squadron Leader N.W. Tucker

1 TORPEDO MAINTENANCE UNIT

CHRONOLOGY

18 November 1942	—formed at Breddan, Qld
May 1944	—headquarters moved to Nowra, NSW
28 June 1944	—Unit disbanded

NARRATIVE

1 Torpedo Maintenance Unit was formed at Breddan near Charters Towers, Queensland, on 18 November 1942, under the command of Flying Officer M. Murray. The Unit also controlled two sections—1 Mobile Torpedo Section at Gurney and 2 Mobile Torpedo Section at Port Moresby.

From December the Unit, despite suffering shortages in equipment, began resupplying 1 Mobile Torpedo Section which was supporting torpedo attacks by the Beauforts of 100 Squadron in New Guinea, with torpedoes and associated stores.

Without any advice to 1 Torpedo Maintenance Unit, 3 Mobile Torpedo Section arrived at Breddan on allocation to the Unit. It was planned to send 3 Mobile Torpedo Section on to Milne Bay, but this was changed to Bohle River, where the section supported 8 Squadron from mid-1943.

Meanwhile the parent unit, in addition to administering its various mobile torpedo sections, which increased in number with the allocation of 4 Mobile Torpedo Section at Higgins Field from July 1943, conducted torpedo loading trials on Catalinas at Cairns.

On 31 July 1943 Flying Officer R.A. Jones replaced Flying Officer Murray as Commanding Officer. The following month, 5 Mobile Torpedo Section, located at Brisbane, came under the control of 1 Torpedo Maintenance Unit and departed for Goodenough Island. 1 Torpedo Maintenance Unit was supporting the torpedo bombing activities of 8 and 100 Beaufort Squadrons and, to a lesser extent, the Catalina flying boats of 11 and 20 Squadrons.

RAAF torpedo bombing operations did not meet with the success expected of them and this resulted in a much reduced use of these expensive and hard-to-maintain weapons from late 1943. As they were no longer required in operational areas, the various mobile torpedo sections re-located to Nowra. 1 Mobile Torpedo Section was the first to move and arrived in October but it was not until April 1944 that the remaining sections returned to Nowra. In May, 1 Torpedo Maintenance Unit Headquarters joined them at Nowra to supervise the wind-up of the various sections. On 28 June 1944 the Unit disbanded.

Commanding Officers

November 1942	—Flying Officer M. Murray
July 1943	—Flying Officer R.A. Jones

2 TORPEDO MAINTENANCE UNIT

CHRONOLOGY

1 February 1943	—formed at Nowra, NSW
24 March 1943	—moved to Adelaide River, NT
May 1943	—last entry in Unit History Record

NARRATIVE

Formed at Nowra under the command of Flying Officer W.T. Smith, 2 Torpedo Maintenance Unit came into existence on 1 February 1943. Aside from provisioning, maintaining and storing torpedoes in its own right the Unit's headquarters was to 'administer such mobile torpedo sections as were assigned to it'. It was initially planned that two mobile torpedo sections would come under the control of the Unit and its establishment was set at four officers, 11 non-commissioned officers and 76 airmen.

After formation, notification was received to move to Adelaide River in the Northern Territory. This re-location commenced on 24 March despite equipment shortages, notably in specialist vehicles, which were to be made good at Adelaide River. On arriving in the 'top end', the Unit was allocated a camp site at Batchelor Airfield, instead of Adelaide River.

By 31 May 1943 the personnel of 2 Torpedo Maintenance Unit were fully concentrated at their new base. Work to complete the camp site was still in progress. The canteen was, however, in operation and sporting events such as inter-unit shuttlecock and tennis competitions were being run to keep personnel occupied after hours.

COMMANDING OFFICER

1 February 1943	—Flying Officer W.T. Smith

AIRCRAFT MAINTENANCE SQUADRON WILLIAMTOWN

CHRONOLOGY

1 February 1987	—formed at RAAF Williamtown, NSW
November 1994	—based at RAAF Williamtown, NSW

NARRATIVE

The Aircraft Maintenance Squadron Williamtown (AIMSWLM) was formed on 1 February 1987 under the command of 481 Wing—a new organisation formed from 481 Squadron, and having Aircraft Maintenance Squadron and Aircraft Equipment Maintenance Squadron under its command. From AIMSWLM's formation until July 1990 no unit history was recorded in the Unit History Record.

The Squadron's role is to perform maintenance on Hornet F404 engines, F/A-18 Hornet, Macchi and Winjeel aircraft. It also runs maintenance training courses related to these aircraft.

The Squadron has run adventure training and bivouacs to Barrington Tops National Park, Myall Lakes National Park, Myall River State Forest and Seal Rocks. Personnel from the Squadron have also been attached to various exercises, for instance 'Pitch Black', 'Stuart Black', 'High Sierra', 'Blue Knight' and 'Churinga' at Butterworth.

The Squadron is still operational and based at RAAF Williamtown.

COMMANDING OFFICERS

1 February 1987	—Wing Commander G. Riley
9 January 1989	—Wing Commander B.W. Fox
February 1991	—Wing Commander R.F. Luke

AIRCRAFT EQUIPMENT MAINTENANCE SQUADRON WILLIAMTOWN

CHRONOLOGY

1 February 1987	—formed at RAAF Williamtown, NSW
November 1994	—based at RAAF Williamtown, NSW

NARRATIVE

The Aircraft Equipment Maintenance Squadron Williamtown (AEMSWLM) was formed on 1 February 1987. On 31 January 1987, 481 Squadron was disbanded and on 1 February 1987 a new organisation was formed—481 Wing with two squadrons. These were Aircraft Equipment Maintenance Squadron and Aircraft Maintenance Squadron. From AEMSWLM's formation until July 1990 no unit history was recorded in the Unit History Record.

The Squadron consists of three flights: the Armament Flight; Avionics Maintenance Flight (consisting of the Avionics Workshops and Automatic Test Equipment Operations); and the Weapons System Support Flight (consisting of the Engineering Support Section and the Software Support Agency). The role of these three flights is to provide maintenance for the Tactical Fighter Group. The Squadron also runs training courses related to maintenance of F/A-18 Hornet aircraft.

Personnel of the Squadron have participated in various exercises including 'Black Night', 'Pitch Black' and 'Blue Knight'. Adventure training has also been held and, in January 1994, members of the Squadron participated with volunteer bushfire fighting forces to protect the base and local communities. At one stage the fire came within 30 meters of the base's fuel farm.

The Squadron is still operational and based at RAAF Williamtown.

COMMANDING OFFICERS

March 1990	—Wing Commander J.W. Grubb
December 1991	—Wing Commander S.B. Lang
February 1994	—Wing Commander P.M. Hewitson

BASE TORPEDO UNIT

CHRONOLOGY

7 September 1942	—formed at RAAF Base Nowra, NSW
31 March 1947	—Unit disbanded

NARRATIVE

By May 1942 the Torpedo Range Section and Torpedo Workshops were segments of the RAAF Nowra organisation, but it was considered more efficient to amalgamate the torpedo sections as a separate unit under the command of RAAF Nowra. On 1 August 1942 the decision was made that RAAF Nowra would administer Base Torpedo Unit as a separate unit, with the latter handling, through Station Headquarters Nowra, torpedo technical, equipment, tactics, training and course syllabuses. Base Torpedo Unit was formally established, under the command of Squadron Leader J.O.H. Dibbs, on 7 September 1942.

The United States Army Air Corps had shown interest in the base at Nowra for torpedo training and on 21 June 1942, four B-26 aircraft under the command of Lieutenants Larson, MacLaren, Lewis and Vroom commenced training; this was completed on 9 July. On 16 July another eight aircraft, under the command of Captain Moye and Lieutenants Hatch, Krell, Green, Craft and Standwood arrived; on 25 July these crews completed training. The Unit also provided personnel to man mobile torpedo sections, the first of which, 1 Mobile Torpedo Section, moved north on 1 August, and was followed by 2 Mobile Torpedo Section on 19 September.

On 1 August 1942 1 Beaufort Course—consisting of the 100 Squadron crews of Wing Commander J.R. Balmer, Flying Officers Green, Stumm and Avery and Sergeant James—commenced torpedo training. A second Beaufort course was undertaken between 17 and 29 August.

Flying training was under the control of Squadron Leader A. Gadd, Chief Flying Instructor and Officer Commanding Torpedo Training Flight, and the Chief Ground Instructor was Lieutenant Drake, United States Navy. The aim was to train 18 crews per month, plus 20 Fitter IIE(T)s every six weeks and Aircraft General Hands (Torpedo) every three weeks.

7 Squadron arrived on 17 September 1942, continuing the training of RAAF torpedo bomber squadrons. Training on these courses consisted of mock attacks on a rescue launch, ship recognition and attacks on moving targets supplied in the form of HMAS *Broome* and HMAS *Tamworth*, among others. In many cases, half the course would be aboard the target ship whilst it was under attack by the other course members before reversing the procedure. Although aircraft serviceability was initially low (it improved as ground crew gained experience with the aircraft) attacks were made in conjunction with the B-25 Mitchell bombers of 18 (NEI) Squadron, to enable realistic training to be undertaken.

Base Torpedo Unit developed a mirror torpedo sight which was successfully tested against the 4000-ton *Montoro* on 6 November. Tactics were also developed. On 21 January 1943 seven aircraft flew to Evans Head, New South Wales, to test tactics to be used during night torpedo attacks. The technique developed consisted of an Air-to-Surface Vessel (ASV) radar-equipped aircraft finding the target, and thence the strike aircraft homing onto this aircraft and making their attack against targets silhouetted by flares dropped from the search aircraft. Squadron Leader Gadd led aircraft on a strike against HMAS *Australia* and escort, USS *Bagley* and USS *Helen*, during February, and on 23 February a torpedo attack was made by a single aircraft using the light of its own flares. The night torpedo technique was tested against *Le Triomphant* on 4 May 1943.

Throughout its active life, the Unit was also involved with dropping trials of Mk XII, Mk XIII and Mk XV torpedoes, as well as undertaking trial dropping of torpedoes from Catalina aircraft during August 1943.

Wartime activity was not without casualties in aircraft and men. On 25 September 1942, Flight Lieutenant Paul of 7 Squadron force-landed Beaufort A9-109 near Bowen Island and the aircraft broke up. Whilst returning from an attack on HMAS *Tamworth* on 29 November 1942, the starboard propeller of A9-105 flew off and Pilot Officer Swanson successfully flew back to Nowra on one engine, only to crash and burn the aircraft when attempting to land. On 3 April, Flying Officer Walsh crashed A9-210 into the sea; A9-96 dipped its port propeller whilst low flying over the water; and A9-101 suffered a seagull strike which shattered the nose perspex. Two days later, Sergeant Parker in A9-117 collided with trees along the shoreline, badly damaging the aircraft, and on 12 April Sergeant Jackson was killed when A9-266 crashed in Cabbage Tree Creek.

The most infamous accident occurred on 14 April 1943. Flight Lieutenant D.G. Ley and Flying Officer R.S. Green and crews were killed in a collision over Jervis Bay whilst carrying out a formation demonstration for war correspondents. The incident was filmed, and featured in the television documentary *This Fabulous Century*.

During December 1944, preparations began for the transfer of the RAAF base at Nowra to the Royal Navy Fleet Air Arm. Base Torpedo Unit co-operated with the Royal Navy during its torpedo dropping exercises, before disbanding on 31 March 1947.

Commanding Officers

7 September 1942	—Squadron Leader J.O.H. Dibbs
7 July 1943	—Wing Commander G.F. Ell
12 April 1944	—Squadron Leader G.L. Grendon
10 May 1944	—Flight Lieutenant A.H. Campey

MAINTENANCE SQUADRON EAST SALE

Ever Sustaining

CHRONOLOGY

1 May 1952	—raised at RAAF East Sale, Vic.
November 1994	—based at East Sale, Vic.

NARRATIVE

On 1 May 1952, Maintenance Squadron East Sale was formed from the technical component of RAAF Station East Sale, with Wing Commander A.J. Abicair as its first Commanding Officer. At the time of its formation the unit was responsible for the maintenance of nine different aircraft types: the Lincoln, Dakota, Mustang, Wirraway, Vampire, Beaufighter, Tiger Moth, Mosquito and Auster.

During its early period of operations the unit faced many difficulties. The initial unit strength of six officers, 251 airmen and two WRAAFs was well below the establishment of 10 officers, 313 airmen and five WRAAFs. This, combined with the operation of the 10 aircraft types and at times severe weather conditions, severely strained resources.

1954 saw the introduction of a new internal organisation considered to best meet the demands placed upon the Squadron. The unit was divided into sections of heavy aircraft, light aircraft, armament and workshops, ancillary section, tarmac, records and ground equipment support maintenance. Notwithstanding continuing poor working and weather conditions, 1954 saw a post-war record flying effort achieved with 11 916 hours being flown on all aircraft types. This was followed the next year by a unit record monthly maintenance effort of 2240 engine flying hours in conjunction with 990 airframe flying hours. 1954 also marked the establishment of the jet training flight.

1955 saw the move of Tiger Moth flying from East Sale to its satellite base at West Sale. Although the aircraft only flew every second day to allow for aircraft repairs, all the 'petrol, oil, spares and handling equipment' had to be transported to the West Sale aerodrome from East Sale. September of the same year saw the introduction of Winjeel aircraft into service at East Sale.

During 1967 and 1968 the ageing Vampire was replaced by the Macchi Trainer. The first loss was recorded on 7 March 1968, when A7-003 was destroyed in an aircraft accident. Also during 1968 the first HS748 Navigation Trainer, A10-601, was received into the RAAF at East Sale as work commenced on two hangars in which they would be serviced. Whilst Maintenance Squadron had, since its inception, serviced ground support equipment, it was not until 1969 that an authority was received to transfer mechanical transport maintenance from Base Squadron.

During April 1971, on the direction of the Department of Air, a new organisation of Maintenance Squadron was implemented. The Squadron was divided into flights and sections: Maintenance Support Flight composed of MCS and Equipment; Field Training Section composed of Training Control Office, Technical Training School and Trade Test Office; Ground Equipment Maintenance Flight composed of Ground Support Equipment Maintenance, Motor Transport Maintenance and General Engineering; Aircraft Maintenance Flight composed of 'A' Flight Line (Macchi, Dakota, Winjeel), Macchi Maintenance, Dakota Maintenance and Winjeel Maintenance Sections, 'B' Flight Line (HS748) and HS748 Maintenance; Aircraft Equipment Maintenance Flight composed of Safety Equipment, Radio, Electrical, Instrument and Armament Sections; and Administration Flight composed of the orderly room and the disciplinary office.

In 1979 Maintenance Squadron East Sale supported Papua New Guinea's Independence Celebration, and the Winjeel aircraft departed from the base for the last time. Replaced by the CT-4 Air Trainer (Plastic Parrot), the Winjeel had served admirably as the primary flying trainer over many years. In November 1979 Macchi flight line personnel vacated their World War II flight line building and took up residence in temporary accommodation. This enabled the old flight line building to be moved from the Macchi hangar area to a temporary position in front of the Dakota hangar. Early in 1980, the Dakota aircraft departed the base for the last time.

In 1986 Hawker de Havilland was awarded the initial maintenance contract on the recently delivered PC-9 Trainer and a milestone was reached—the 10 000th flying hour for a RAAF HS748. Four years later, the HS748 Navigation Trainer aircraft fleet maintained by the Squadron recorded a total of 100 000 accident-free flying hours.

By October 1988 the new headquarters of the Squadron was completed. Six weeks later the Squadron was presented with the RAAF Maintenance Trophy for the 1987–88 year. The highlight of 1989 was the presentation of the Governor-General's Banner to the Squadron, given in recognition of 25 years outstanding service by a non-operational squadron.

1 January 1991 marked the expiry of the civilian contract for PC-9 servicing and 16 of the civilians filled newly created positions at HS748 Section. By the end of June, flying hours for the 1990–91 financial year totalled 13 764. On 10 December, all 10 CT-4 aircraft departed the base for the last time.

With a unit strength of 220, the Squadron provides and maintains 16 PC-9 aircraft to Central Flying School. Ten HS748 aircraft are maintained for use by 32 Squadron. While the unit fully maintains a large number of avionic items, the more complex communication items are despatched to centralised facilities for repair and overhaul. Responsibility is also held for the servicing of all motor transport vehicles and ground support equipment in use on base. In addition, Squadron personnel maintain flying safety equipment, provide fuel quality control services for the fuel installations and service the fuel tankers in use. The Squadron also has responsibility for the calibration of all test equipment used on the base.

Commanding Officers

1 May 1952	—Wing Commander A.J. Abicair
February 1954	—Wing Commander I.McK. Keir
2 May 1957	—Wing Commander C.W. Butcher
24 March 1961	—Wing Commander K.P. Connolly
9 December 1963	—Wing Commander R.E. Cormie
September 1967	—Wing Commander C.M. Bevan
19 January 1970	—Wing Commander J.F. Flannery

1 September 1970	—Wing Commander J.A. Dietz
10 January 1972	—Wing Commander B.J. Graf
12 February 1974	—Wing Commander R.J.H. Meredith
13 February 1976	—Wing Commander R.H. Tucker
11 January 1978	—Wing Commander C.G. Beatty
February 1980	—Wing Commander R.C. Newbery
February 1982	—Wing Commander G.T. Leo
September 1983	—Wing Commander B. White
January 1986	—Wing Commander J.W. Baker
11 January 1988	—Wing Commander G.F. Bates
11 May 1990	—Wing Commander J.W. Clarkson
14 December 1992	—Wing Commander P.W.T. McCormack
February 1994	—Squadron Leader L.R. Brown

MAINTENANCE SQUADRON EDINBURGH

CHRONOLOGY

31 October 1955	—formed at Edinburgh, SA
1 June 1965	—Squadron disbanded

NARRATIVE

Maintenance Squadron Edinburgh was formed on 31 October 1955, under the temporary command of Squadron Leader J.H. Gooch, with a total strength of 152 officers and airmen. The headquarters took up residence in a building in the hangar area on 11 January 1956, and the Squadron settled into its final organisation with the raising of an ancillary flight, composed of radar, electrical and instrument sections, on 12 September 1956. Ancillary Flight was commanded by Squadron Leader Gooch, who relinquished command of the Squadron to Wing Commander J.C. Kane on 13 February 1956.

During its period of service, Maintenance Squadron Edinburgh serviced and modified a multitude of aircraft types: Dakota, Bristol Freighter, Lincoln, Beaufighter, Canberra, Mustang, Sabre, Winjeel, Valiant (for special modifications), Beaver, Otter, Sea Vixen, Pika (the manned prototype of the Jindivik target aircraft) and the Jindivik. In addition, in co-operation with Fairey Aviation, the Squadron converted Meteor and Canberra aircraft into radio-controlled drones.

Much of the work was undertaken for the Air Trials Unit at Woomera.

It was not only in the aircraft maintenance field that the unit excelled. On 2 March 1961, Flying Officer R.A. Dunnett proceeded to Woomera to undertake testing of gear to be used in conjunction with the Operation 'Mercury' (the first United States astronaut program) station at Muchea, Western Australia. Dunnett worked closely with the program from 18 to 20 April 1961 and between 19 and 23 June 1961. The Squadron was also involved with the setting up of a Dakota aircraft with essential equipment for the 'Mercury' program.

With the commencement of 24 Squadron (Citizen Air Force) camps of continuous training and training weekends, members of the radio, electrical and instrument sections acted as lecturers. These weekends were assessed by the executives of 24 Squadron as being very successful.

In common with all service units, the Squadron supplied personnel to contribute to various guards of honour. On one such occasion Flight Lieutenant Furler led a guard of honour for United States General Doolittle on 12 May 1956 and 24 members contributed to the guard of honour at the Adelaide Airport when the Sir Ross and Sir Keith Smith Memorial was unveiled on 27 April 1958—a fitting reward for assistance given to the Vickers Vimy project.

Six members of the Squadron were witnesses to the atomic bomb tests at Maralinga on 4 October 1956 and 9 November 1957.

Maintenance Squadron Edinburgh also had a fine record in the sporting field. Members of the Squadron were individually, and as team members, prominent in swimming, athletics, rifle shooting, tennis, table tennis and badminton.

The unit disbanded on 1 June 1965, its function being passed to 2 Air Trials Unit. When the transfer was made, the aircraft which were under its charge for maintenance consisted of one Bristol Freighter, two Canberra B Mk 2, six Meteor U Mk 21, one Canberra U Mk 10, one Jindivik Mk 3A and a single Meteor T7. This line-up of aircraft being serviced is typical of the numbers and diversity of types which were handled by Maintenance Squadron Edinburgh during its period of activity.

Commanding Officers

13 February 1956	—Wing Commander J.C. Kane
30 April 1956	—Squadron Leader S.C. Medley
3 February 1958	—Wing Commander C.J. Leopold
9 July 1962	—Wing Commander L.S.W. Smyth
28 January 1964	—Wing Commander W.V. Greenham

MAINTENANCE SQUADRON POINT COOK

CHRONOLOGY

14 June 1950 —raised at RAAF Base Point Cook, Vic.
2 June 1958 —Squadron disbanded

NARRATIVE

Maintenance Squadron Point Cook came into existence when the Technical Squadron, RAAF Station Point Cook, was disbanded. Originally known as 25 Maintenance Squadron, Maintenance Squadron Point Cook was initially commanded by Wing Commander P.S. Kennedy.

The Squadron was tasked with the maintenance of aircraft operating from Point Cook, which included such varied types as Oxford, Tiger Moth, Wirraway and Winjeel aircraft. To be successful in this field, the Squadron constantly upgraded its expertise and initiative. For example, the Squadron Armament Section designed and built prototype gun mounts to enable two .303 machine guns to be mounted on Wirraway aircraft.

The Squadron also undertook salvage tasks. On the first of these a party travelled to Lake Corangamite between 20 October and 3 November 1950 to attempt to salvage a Wirraway which had crashed into the lake. Another Wirraway was recovered near Mortlake, Victoria, on 22 June 1953; on the next day the same type of aircraft was recovered after it force-landed near Illabarook, Victoria. During the period 17–18 August, a Wirraway was partially salvaged after crashing near Wallan.

The Squadron also assisted with the maintenance of aircraft operating out of Point Cook. For example, 29 aircraft from 75, 76 and 21 Squadrons arrived on 11 September 1951 to participate in the Laverton Air Pageant, which took place on the 15th. The aircraft were serviced by the staff of Maintenance Squadron Point Cook, thus enabling them to participate in the pageant.

There were also unusual aircraft, for unusual tasks. On 13 November 1953, two Auster aircraft arrived for Exercise 'Frigid Flutter'. In this exercise the Austers were to operate as an integral part of an Australian Antarctic expedition to establish a base on the Antarctic continent.

Members of the staff were given lectures on various aspects of their trades—such as North American (as fitted to the Sabre) and Martin Baker ejection seats—and encouraged to study electronics and the like at civilian institutions in Melbourne.

In addition to its technical tasks, Maintenance Squadron Point Cook was often requested to participate in formal guards and ceremonies. Members of the Squadron took pride of place during the Point Cook Coronation Parade on 21 June 1953, and when Queen Elizabeth II visited Point Cook on 6 March 1954 60 members of the Squadron were members of the escort wing for the Colours. Two months later, the Squadron formed a contingent of the Coral Sea Parade and, on 18 May 1955, provided a guard of honour at the Australian Flying Corps Memorial at Point Cook.

Maintenance Squadron Point Cook was disbanded on 2 June 1958.

Commanding Officers

14 June 1950	—Wing Commander P.S. Kennedy
16 April 1956	—Wing Commander E.K. De Cean
9 December 1957	—Squadron Leader H.D. Paull

RADIO INSTALLATION AND MAINTENANCE UNITS

Narrative

The six radio installation and maintenance units were formed to install and maintain radio, radar and telecommunication equipment. In particular, the units undertook the calibration of radio direction finding equipment, testing of radar beacon equipment, the installation of towers, arrays and transmitting stations, modification and testing of Identification Friend or Foe (IFF), Air-to-Surface Vessel (ASV) radar equipment, and High Frequency (HF) radio installation. The units' area of operations covered a large area of Australia and the South-West Pacific.

Of all of the units, 1 was the largest and served for the longest period. It was formed at Croydon, New South Wales, and occupied the Presbyterian Ladies College in July 1942. It carried out operational work in New South Wales and, to a lesser degree, Queensland and Victoria. Due to the nature of work carried out in this Unit, in January 1945, the name of the Unit was changed to the Radio Development and Installation Unit and, in March 1945, a Melbourne detachment was formed. On 17 September 1945 the Croydon headquarters was closed but reopened on 24 September 1945 at Somers, Victoria. This included the Melbourne detachment. In July 1946 the Unit became the Care and Maintenance Unit and proceeded to disband. The final disbandment took place at 4 Maintenance Group Headquarters.

2 Radio Installation and Maintenance Unit was formed in October 1944 at Townsville and while still operational in December 1945, was in the process of disbandment. 5 was formed in August 1944 at Darwin, from the disbanded 11 Signals Unit and 44 Radar Wing. It was involved in HF, VHF, radar and land-line communications in the Northern Territory and finally disbanded on 1 March 1946.

The other three units were based in the South-West Pacific Area. 3 was formed at Milne Bay in July 1944, after the disbandment of 10 Signals Unit. The Unit moved to Madang and then on to Kranket Island, where it ceased to function on 31 January 1946.

4 formed on 1 July 1944 in New Guinea from the disbanded 13 Signals Unit, and served at Noemfoor, Morotai and Labuan, where it closed on

7 December 1945. Though the Unit closed, a nucleus moved to Kingaroy, Queensland, and the Unit officially disbanded in May 1946.

6 was formed on 18 September 1945 at Morotai from a detachment of 4, and only existed until November, when work commenced on demolition of the camp site in preparation for final disbandment.

1 Radio Installation and Maintenance Unit

Chronology

3 June 1942 —formed at Croydon, NSW
15 January 1945 —re-named Radio Development and Installation Unit
31 March 1945 —Melbourne detachment formed
24 September 1945 —Croydon and Melbourne detachments moved to Somers, Vic.
July 1946 —re-named Care and Maintenance Unit
7 November 1946 —Unit disbanded

Commanding Officers

3 June 1942 —Squadron Leader A.E. Mitchell
14 August 1942 —Wing Commander H.S. Preston
25 August 1944 —Squadron Leader B.F.N. Israel
8 December 1944 —Squadron Leader J.C. Norrie
13 December 1945 —Wing Commander J.E. Reynolds
15 July 1946 —Flight Lieutenant G.C. Zucker

2 Radio Installation and Maintenance Unit

Chronology

1 October 1944 —formed at Townsville, Qld
31 December 1945 —last entry in Unit History Record

Commanding Officer

1 October 1944 —Squadron Leader A.E. Walz

3 Radio Installation and Maintenance Unit

Chronology

1 July 1944 —formed at Milne Bay
31 January 1946 —ceased to function

Commanding Officers

1 July 1944 —Wing Commander S.A.F. Westwood
7 July 1944 —Squadron Leader A.C. Gray (RAF)
18 January 1945 —Squadron Leader W.W. Weston
12 November 1945 —Squadron Leader S.G. Edwards

4 Radio Installation and Maintenance Unit

Chronology

1 July 1944 —formed in New Guinea
7 December 1945 —closed at Labuan, North Borneo
2 January 1946 —nucleus at Kingaroy, Qld
28 May 1946 —Unit disbanded

Commanding Officers

1 July 1944 —Squadron Leader J.M. Brennan
11 July 1944 —Squadron Leader A.M.J. Clark
14 August 1945 —Squadron Leader Curtain
25 August 1945 —Squadron Leader Stanfield
2 January 1946 —Flight Lieutenant L.G. Hall

5 Radio Installation and Maintenance Unit

Chronology

22 August 1944 —formed in Darwin, NT
1 March 1946 —Unit disbanded

Commanding Officers

22 August 1944 —Squadron Leader J.E. Reynolds
19 October 1945 —Squadron Leader G.W. Marriott

6 Radio Installation and Maintenance Unit

Chronology

18 September 1945 —formed on Morotai
30 November 1945 —prepared for disbandment

Commanding Officer

18 September 1945 —Wing Commander A.C.A. McBride